Enhanced Biological Phosphorus Rem

Metabolic Insights and Salinity Effec

Enhanced Biological Phosphorus Removal
Metabolic Insights and Salinity Effects

Submitted in fulfillment of the requirements of

the Board for Doctorates of Delft University of Technology

and
of the Academic Board of the UNESCO-IHE

Institute for Water Education

for
the Degree of DOCTOR

to be defended in public on

Tuesday, 22nd December 2015, at 15:00 hours

In Delft, the Netherlands

by

Laurens WELLES

Master of Science, Delft University of Technology

born in The Hague, the Netherlands.

This dissertation has been approved by the promotors:
Prof. dr. D. Brdjanovic
Prof. dr. ir. M.C.M. van Loosdrecht

Composition of Doctoral Committee:

Chairman	Rector Magnificus TU Delft
Vice-Chairman	Rector UNESCO-IHE
Prof. dr. D. Brdjanovic	UNESCO-IHE/Delft University of Technology, promotor
Prof. dr. ir. M.C.M. van Loosdrecht	Delft University of Technology, promotor

Independent members:

Prof. dr. J.G. Kuenen	Delft University of Technology
Prof. dr. G.A. Ekama	University of Cape Town, Cape Town, South Africa
Prof. dr. P.H. Nielsen	Aalborg University, Aalborg, Denmark
Prof. dr. G.H. Chen	Hong Kong University of Science and Technology, Clear Water Bay, Hong Kong
Prof. dr. M.D. Kennedy	UNESCO-IHE/Delft University of Technology, reserve

CRC Press/Balkema is an imprint of the Taylor & Francis Group, an informa business

Published by:
CRC Press/Balkema
PO Box 11320, 2301 EH Leiden, The Netherlands
e-mail: Pub.NL@taylorandfrancis.com
www.crcpress.com – www.taylorandfrancis.com

ISBN 978-1-138-02947-7 (Taylor & Francis Group)

To
Jahmahn
and
Elias

Table of contents

Abbreviations and Symbols

Abbreviations

AOB: Ammonium oxidizing bacteria
ADP: Adenosine diphosphate
ATP: Adenosine triphosphate
ATU: Allyl-N-thiourea
BOD: Biochemical oxygen demand
BPR: Biological phosphorus removal
C: Carbon
COD: Chemical oxygen demand
DGGE: Denaturing gradient gel electrophoresis
DO: Dissolved Oxygen
DPAO: denitrifying Polyphosphate-accumulating Organisms
EBPR: Enhanced biological phosphorus removal
ED: Entner-Doudoroff
EMP: Embden-Meyerhof-Parnass
FISH: Fluorescence *in situ* hybridization
GAM: Glycogen-accumulating organisms metabolism
GAO: Glycogen-accumulating organisms
Gly: Glycogen
HAc: Acetate
HPr: Propionate
HRT: Hydraulic retention time
ISS: Inorganic suspended solids
MAR: Microautoradiography
MLSS: Mixed liquor suspended solids
MLVSS: Mixed liquor volatile suspended solids
MUCT: Modified University of Cape Town (configuration)
N: Nitrogen
NAD+: Nicotinamide adenine dinucleotide
NADH: Reduced form of nicotinamide adenine dinucleotide
NADP+: Nicotinamide adenine dinucleotide phosphate
NADPH: Reduced form of nicotinamide adenine dinucleotide phosphate
NOB: Nitrite oxidizing bacteria
OHO: ordinary heterotrophic organisms
OUR: Oxygen uptake rate
P: Phosphorus
PAM: Polyphosphate-accumulating organisms' metabolism
PAO: Polyphosphate-accumulating organisms
PAO I: *'Candidatus* Accumulibacter phosphatis' clade I
PAO II: *'Candidatus* Accumulibacter phosphatis' clade II
PCR: Polymerase chain reaction
PHA: Poly-β-hydroxyalkanoates
PHB: Poly-β-hydroxybutyrate
PHV: Poly-β-hydroxyvalerate
PH2MV: Poly-β-hydroxy-2-methyl-valerate
PH2MB: Poly-β-hydroxy-2-methylbutyrate
Poly-P: Poly-phosphate
SBR: Sequencing batch reactor
SRB: Sulphate reducing bacteria
SRT: Solids retention time
T: Temperature
TSS: Total suspended solids
VFA: Volatile fatty acids

VSS: Volatile suspended solids
WWTP: Wastewater treatment plant

Symbols

Chapter 2, 3, 4 and 5

ISS_b: Inorganic suspended solids associated with active biomass
ISS_{pp}: Inorganic suspended solids associated with Poly-P
$T_{P,i}$: Total phosphorus concentration in the influent
$T_{P,e}$: Total phosphorus concentration in the effluent
$f_{P,TSS}$: Ratio of total P per TSS
$f_{P,ppVSS}$: Ratio of Poly-P per VSS
$f_{P,bVSS}$: Ratio of non Poly-P phosphorus per VSS
$f_{P,VSS}$: Ratio of total P per VSS
$f_{ISSb,TSS}$: Ash content associated with active biomass
$f_{P,ppISS}$: P-content of poly-P
V_p: Working volume of reactor
Q_i: Influent flow rate
Q_w: Wastage of biomass flow rate
Pns: Non-soluble total phosphorus
P_b: Phosphate associated with active biomass
$q_{P,ana}^{MAX}$: Maximum active biomass specific anaerobic P-release rate
$q_{SA,ana}^{MAX}$: Maximum active biomass specific anaerobic VFA-uptake rate
$q_{P,aer}^{MAX}$: Maximum active biomass specific aerobic P-uptake rate

Chapter 6

$m_{ATP,PAO\_total}^{an}(S)$: Anaerobic PAO Maintenance coefficient at different salinity concentrations
$m_{ATP,PAO\_poly-P}^{an}(S)$: Anaerobic PAO poly-P maintenance coefficient at different salinity concentrations
$m_{ATP,PAO\_gly}^{an}(S)$: Anaerobic PAO glycogen maintenance coefficient at different salinity concentrations
$m_{ATP,GAO}^{an}(S)$: Anaerobic GAO Maintenance coefficient at different salinity concentrations
$m_{ATP}^{an,re}(S)$: Anaerobic required maintenance coefficient at different salinity concentrations
$f_{i,1}(S)$: Empirical inhibition factor at different salinity concentrations
S: Salinity concentration
m_{ATP}^{0} : Maintenance coefficient at 0% salinity concentration
a: Linear proportional increase in maintenance requirements per increase in salinity
bi_1: Impact factor, describing the magnitude of the inhibition effect
Si_1: Salinity concentration at which 50% inhibition occurs

$q_{SA,PAO\_total}^{MAX}(S)$: Total maximum PAO acetate uptake rate at different salinity
$q_{SA,PAO\_PAM}^{MAX}(S)$: Maximum PAO acetate uptake rate facilitated by a PAM at different salinity
$q_{SA,PAO\_GAM}^{MAX}(S)$: Maximum PAO acetate uptake rate facilitated by a GAM at different salinity
$q_{SA,GAO}^{MAX}(S)$: Maximum GAO acetate uptake rate at different salinity
$q_{P,PAO\_HAc}^{MAX}(S)$: Maximum PAO PO$_4$ release rate at different salinity
$f_{i,2}(S)$: Empirical inhibition factor, describing the inhibition on the PAO acetate uptake and P-release facilitated by a PAM and the GAO acetate uptake at different salinity
$f_{a,1}(S)$: Empirical activation factor, describing the activation of the PAO acetate uptake facilitated by a GAM at different salinity
$q_{SA,PAO\_PAM}^{MAX,0}$: Maximum PAO acetate uptake rate facilitated by a PAM at 0% salinity
$q_{SA,PAO\_GAM}^{MAX,0}$: Maximum PAO acetate uptake rate facilitated by a GAM at 0% salinity
$q_{SA,GAO}^{MAX,0}$: Maximum GAO acetate uptake rate at 0% salinity

$q_{P,PAO\_HAc}^{MAX,0}$: Maximum PAO PO$_4$ release rate at 0% salinity

bi_2: Impact factor, describing the magnitude of the inhibition effect on the acetate uptake and P-release effect

Si_2: Salinity concentration at which 50% inhibition of the acetate uptake occurs

ba_1: Impact factor (equal to bi_2), describing the magnitude of the activation effect on the acetate uptake

Sa_1: Salinity concentration (equal to Si_2) at which 50% activation of the PAO GAM acetate uptake occurs.

$f_{P/HAc}^{total}$: Total P-release rate /HAc uptake rate at different salinity

$f_{P/HAc}^{HAc}$: P-release rate corrected for maintenance acitvity/HAc uptake rate at different salinity

$f_{\Delta gly/\Delta HAc}^{total}$: Net glycogen consumption / net HAc uptake

$f_{\Delta gly/\Delta HAc}^{HAc}$: Net glycogen consumption corrected for maintenance glycogen consumption / net HAc uptake

$f_{gly/HAc}^{total}$: Glycogen consumption rate /HAc uptake rate

$f_{gly/HAc}^{HAc}$: Glycogen consumption rate corrected for the maintenance glycogen consumption rate/HAc uptake rate

ΔHAc: Net acetate consumption during the tests

Δt: Time interval of batch test

X_{GAO}: Active biomass concentration of GAO

Chapter 7

$m_{ATP,PAO_{total}}^{o}(S)$ PAO total aerobic maintenance coefficient at different salinity concentrations

$m_{ATP,PAO_{O2}}^{o}(S)$: PAO aerobic maintenance energy generated by O$_2$ consumption at different salinity concentrations

$m_{ATP,PAO_{poly-P}}^{o}(S)$ PAO aerobic maintenance energy generated by poly-P consumption at different salinity concentrations S: Salinity concentration

$m_{ATP,1}^{o,0}$: O$_2$ maintenance coefficient at 0% salinity concentration

a_1: O$_2$ linear proportional increase in maintenance requirements per increase in salinity

bi_1: Impact factor, describing the magnitude of the inhibition effect on O$_2$ maintenance

Si_1: Inhibition factor, determining in which salinity range the inhibition starts to occur of the O$_2$ maintenance

$q_{O2,PAO\_cor}^{o}(S)$: PAO O$_2$ uptake rate at different salinity, corrected for maintenance O$_2$ consumption

$q_{PO4,PAO\_cor}^{o}(S)$: PAO PO$_4$ uptake rate at different salinity, corrected for maintenance P-release

$q_{NH4,PAO}^{o}(S)$: PAO NH$_4$ uptake rate at different salinity

$q_{O2,PAO\_P,N,Gly}^{o,0}$: PAO O$_2$ uptake rate at 0% salinity for PO$_4$, NH$_4$ and glycogen

bi_2: Impact factor, describing the magnitude of the inhibition effect on the PO$_4$ uptake rate

Si_2: Salinity concentration at which 50% inhibition of the PO$_4$ uptake rate occurs

$q_{O2,PAO\_res}^{o,0}$: PAO residual O$_2$ uptake rate

$q_{PO4,PAO\_cor}^{o,0}$: PAO PO$_4$ uptake rate at 0% salinity corrected for PO$_4$ maintenance

bi_3: Impact factor, describing the magnitude of the inhibition effect on the PO$_4$ uptake rate

Si_3: Salinity concentration at which 50% inhibition of the PO$_4$ uptake rate occurs

$q_{NH4,PAO}^{o,0}$: PAO NH$_4$ uptake rate at 0% salinity

bi_4: Impact factor, describing the magnitude of the inhibition effect on the NH$_4$ uptake rate

Si_4: Salinity concentration at which 50% inhibition of the NH$_4$ uptake rate occurs

Summary

Enhanced biological phosphorus removal (EBPR) is a biological process for efficient phosphate removal from wastewaters through intracellular storage of polyphosphate by polyphosphate-accumulating organisms (PAO) and subsequent removal of PAO from the system through wastage of sludge. In comparison to physical and chemical phosphorus removal processes, the biological process has several advantages such as high removal efficiency, low cost, and no chemical sludge production, but disturbances and prolonged periods of insufficient phosphate removal are still observed in conventional treatment systems and the applicability of the process for the treatment of saline waters remains unclear. In this PhD project two different aspects of the enhanced biological phosphorus removal were studied. In the first part of the research, potential existence of functional diversity among PAO clades and its influence on process performance was investigated, whereas in the second part of the study, salinity effects were assessed on the metabolism of polyphosphate-accumulating organisms (PAO) and glycogen-accumulating organisms (GAO).

Functional diversity among PAO clades (Chapter 2 to 5)

Although genetic diversity among PAO clades has been observed in past studies, PAO were often considered to behave functionally the same. Several recent studies suggested that PAO clades may be functionally different and that some PAO clades can perform better phosphate removal than other clades. Considering the significant role of the EBPR process in nutrient removal and recovery processes and the potential effect of PAO clades prevalence on the process performance, there was a need to investigate the potential existence of functional differences among PAO clades. The objective of this part of the study was to assess the existence of functional differences among PAO clades regarding the anaerobic metabolism in relation to their storage polymers and regarding the denitrification pathways.

In **Chapter 2**, it was demonstrated in short-term experiments that significant functional differences exist between PAO I and II, with respect to the anaerobic volatile fatty acid (VFA) uptake metabolism. Although both PAO clades were able to shift their metabolism from a mixed poly-P and glycogen dependent metabolism to a metabolism that fully relies on glycogen, the HAc-uptake rate of both PAO clades decreased significantly where the decrease of HAc-uptake rates was most pronounced for PAO I. Consequently, at poly-P depleted conditions, the HAc-uptake rate of PAO II was four times faster than that of PAO I, whereas that of PAO I was slightly faster at poly-P non-limiting conditions. In addition, under conditions where poly-P was not limiting for the anaerobic HAc-uptake, PAO II performed a mixed metabolism that was partially dependent on glycogen and partially on poly-P for the generation of energy required of the HAc uptake, while PAO I relied to a much bigger extent on poly-P for the generation of energy. These findings are of major importance because they contribute to explain and clarify the controversy concerning the different stoichiometric and kinetic values observed in EBPR systems and are relevant for the development of operational guidelines for combined chemical and biological phosphate removal processes.

In **Chapter 3**, the effect of the storage polymers on the metabolism of PAO II was assessed in long-term experiments and the results were compared to a previous study which was, based on the reported stoichiometry and kinetics, presumably conducted with a PAO I dominated biomass culture. The study supported the observations in short-term experiments regarding the functional diversity between PAO I and II. In addition, it provided interesting insights in the role of storage polymers on the regulation of the anaerobic HAc-uptake metabolism. As the influent P/C ratio increased, the poly-P content of the biomass increased while its glycogen content decreased. At higher P-contents, the kinetic P-release rates for HAc-uptake and maintenance increased. In parallel, the HAc-uptake rates increased up to an optimal poly-P/glycogen ratio of 0.3 P-mol/C-mol. Above that optimal ratio, the HAc-uptake rate decreased. The stoichiometry of the anaerobic conversions showed that a metabolic shift occurred from a glycogen dependent metabolism towards a poly-P dependent metabolism when the poly-P content of the biomass increased. The changes in the HAc-uptake rates suggest that at low poly-P contents the ATP formation rate is the rate limiting step, while at high P-contents (and, thus, low glycogen contents) the NADH production rate becomes the rate limiting step for HAc-uptake. Electron microscopy showed that poly-P is stored in the form of large granules in each PAO cell and, therefore, the rate of poly-P consumption may be surface area limited. Therefore, a decrease in the poly-P content of the biomass could limit the ATP production and thereby trigger the ATP production from glycogen conversion at a smaller rate. The findings contribute to a better understanding of the *Accumulibacter* clades metabolism under dynamic conditions and clarify population dynamics observed in previous studies.

To confirm the observations in Chapter 2 and 3, it was assessed in **Chapter 4** if certain PAO clades had the ability to proliferate under conditions where the phosphate concentrations were just enough for assimilation into biomass. In a SBR system, inoculated with activated sludge, a mixed PAO-GAO culture was enriched after 16 SRT that comprised of 49% PAO and 46% GAO of the total bacterial population. More specifically, all PAO were closely related to '*Candidatus* Accumulibacter phosphatis' Clade II. Under anaerobic conditions, the mixed PAO-GAO culture performed a typical GAO metabolism in which all energy for HAc-uptake was produced by the conversion of glycogen. This study confirmed the findings in chapter 2 and 3 that PAO in general can perform a glycogen dependent metabolism but that PAO II had a competitive advantage over PAO I under phosphate limiting conditions. Under aerobic conditions PAO II were capable of instantly taking up excessive amounts of phosphate when additional phosphate was added to the reactor. The study also demonstrated that from a practical perspective, PAO may remain in PAO II dominated activated sludge systems under phosphate limiting conditions for periods of up to 16 SRT for instance due to overdosing of iron while still being able to take up phosphate aerobically when phosphate becomes available in the influent.

Chapter 5 focussed on the denitrification pathways of PAO. Several literature studies suggest that PAO I is able to use both nitrite and nitrate as external electron acceptor while PAO II is only able to use nitrite. The results from those previous studies are contradictory and inconclusive as no studies were conducted with EBPR cultures highly enriched with specific PAO clades under appropriate conditions. In chapter 5 the oxidative pathways (oxygen, nitrite and nitrate) of a PAO I culture were investigated in combination with different VFA feed (HAc and HPr), firstly after a cultivation period in anaerobic/anoxic mode and secondly after a cultivation period in anaerobic/anoxic/oxic mode. After cultivation in anaerobic/anoxic/oxic mode, the enriched culture was not able to take up P in the presence of nitrate, despite the observation of low denitrification rates. In the presence of oxygen and nitrite, rapid P-uptake was observed. The big difference in denitrification rates with nitrite and nitrate together with observation that side populations were still present in the highly enriched biomass, resulted in the hypothesis that the side population in the biomass might have been responsible for NO_3 to NO_2 conversions, where the carbon source was mainly obtained from released soluble microbial products. This hypothesis was further supported by a comparison of literature values from studies conducted with PAO enrichment with various degrees of PAO enrichment. This comparison showed that the biomass specific P-uptake rate in the presence of nitrate increases when the fraction of side populations increases. In addition, the P-removal/N-removal ratio in many past EBPR studies under anaerobic/anoxic/oxic conditions with nitrite was higher than the P-removal/N-removal with nitrate, suggesting that PAO in general are not capable of using nitrate as external electron acceptor and are dependent on the partial denitrification activity of other organisms.

Overall, this research revealed that significant functional diversity exist in the metabolism of PAO regarding the anaerobic metabolism while the study suggest that for the denitrification pathways among the *Accumulibacter* clades PAO I and II, functional differences may not exist. The differences in the anaerobic metabolism contribute to a better understanding of metabolic differences observed in past studies, provides more insight in population dynamics and are from a practical perspective in particular relevant for the development of nutrient recovery and/or combined chemical and biological P-removal systems. The findings of the denitrification pathways of PAO I provide a better understanding of the role of PAO in combined nutrient removal systems, helps to explain practical issues such as anoxic P-release in full scale wastewater treatment plants and support the development of measures to mitigate such issues in the performance. In addition to the functional diversity, the research provided more insight in the role of the storage polymers on the regulation of the anaerobic substrate uptake metabolism, which also leads to a better understanding of EBPR processes in full scale treatment plants under dynamic conditions. Although this study provided clear insights in the functional diversity of PAO clades and their metabolism, it is just the starting point of research focused on the functional diversity of PAO clades. To enable future research on the functional diversity, reliable selection methods or methods for isolation of PAO should be developed to obtain highly enriched or pure cultures with specific PAO clades.

Impact of salinity on the metabolism of PAO and GAO during short-term exposure (Chapter 6 and 7)

Saline wastewater can be generated by industry, intrusion of saline water in the sewerage or when saline water is used directly as alternative water source for non-potable purposes such as flushing toilets. To prevent the environment from severe environmental issues like hypoxia and eutrophication, the nutrients (C, N and P) need to be removed from saline wastewaters before its discharge to the receiving water bodies. However, salinity may negatively affect the microorganisms responsible for the nutrient removal in biological nutrient removal systems. This study assessed the effect of salinity on the metabolism of the microbial populations that prevail in EBPR systems (PAO and GAO).

In **Chapter 6**, the short-term salinity effects on the anaerobic metabolism of PAO and GAO were assessed. It was demonstrated that salinity affected both PAO and GAO, with PAO being the most sensitive organisms. With increasing salinity the HAc uptake rates were inhibited while the maintenance requirements increased (up to 4% salinity) for both PAO and GAO. Interestingly, elevated salinity levels seemed to induce a shift from poly-P to glycogen consumption for HAc uptake and maintenance by PAO, whereas the stoichiometry of GAO related to the anaerobic HAc-uptake was unaffected. In addition, a structured model was developed, which could successfully describe the salinity effects on the different metabolic processes of PAO and GAO.

In **Chapter 7**, the short-term salinity effects on the aerobic metabolism of PAO were assessed. The metabolism was very sensitive to even low salinity concentrations. An increase from 0.02 to 0.18% salinity led to a decrease in the specific oxygen consumption, PO_4 and NH_4 uptake rates of 25%, 46% and 63%, respectively. At 0.35% and higher salinity concentrations, the PO_4-uptake, NH_4-uptake and glycogen recovery were fully inhibited. Biomass growth was the most inhibited parameter, followed by poly-P formation and glycogen synthesis. The aerobic maintenance energy requirements increased up to a threshold concentration of 2% salinity, above which it rapidly decreased. To supply additional energy to cover the increasing maintenance requirements, P was released at salinity concentrations higher than 0.35%. The aerobic maintenance P-release followed a similar trend like the maintenance oxygen consumption. The inhibition model developed in this study could successfully describe the observed salinity effects on the different metabolic processes in this study.

Overall, the research demonstrated that the EBPR process, in particular the aerobic stage, may be very sensitive to salinity. The findings suggest that any saline discharge equivalent to more than 5% seawater (with 3.4% salinity) addition or 15% brackish water (with 1.2% salinity) by either seawater toilet flushing, industrial discharges or saline intrusion can cause serious upsets of the EBPR process. This indicates that the EBPR process may not be applicable for saline wastewater treatment and that salinity may even be a relevant inhibition factor for activated sludge systems, treating conventional domestic wastewaters. The presented data is however based on short-term (hours) experiments using sodium chloride. To give the organisms the opportunity to acclimatize to higher salinity concentrations or giving the system the possibility to select for more salt tolerant PAO strains, future studies should focus on the long-term salinity effects on EBPR cultures, considering different salt compositions as well as different compositions of organic carbon in the synthetic wastewater.

Samenvatting

Biologische fosfaatverwijdering (EBPR) met fosfaat accumulerende organismen (PAO), is een efficiënt proces voor de verwijdering van fosfaat uit afvalwater. Dit wordt gerealiseerd door middel van de intracellulaire opslag van fosfaat in de vorm van polyfosfaat en daaropvolgende verwijdering van PAO biomassa uit het systeem via het spuislib. In vergelijking tot andere fysische en chemische fosfaatverwijderingsprocessen, heeft het biologische EBPR proces een aantal voordelen zoals hoge verwijderingefficiëntie, lage kosten, en geen chemische slibproductie. Echter, verstoringen van het fosfaatverwijderingsproces en langere perioden met onvoldoende fosfaatverwijdering komen in sommige gevallen nog steeds voor in afvalwaterzuiveringssystemen en bovendien is er onduidelijkheid omtrent de toepasbaarheid van het proces voor de behandeling van zoute afvalwater stromen. In dit promotieonderzoek zijn twee verschillende aspecten van het EBPR proces bestudeerd. In het eerste deel van de studie is er onderzoek gedaan naar functionele verschillen tussen PAO clades in relatie tot de procesprestaties. In het tweede deel van de studie zijn de effecten van verhoogde zoutgehaltes op het metabolisme van PAO en glycogeen accumulerende organismen (GAO) onderzocht.

Functionele diversiteit onder PAO clades (Hoofdstuk 2 tot en met 5).

Ondanks het feit dat de genetische diversiteit onder PAO in eerdere studies waargenomen is, werd er vaak verondersteld dat verschillende clades zich functioneel hetzelfde gedragen. Verscheidene recente studies suggereerden echter dat PAO clades zich mogelijk verschillend gedragen en dat sommige PAO clades fosfaatverwijdering beter kunnen uitvoeren dan andere clades. Door het potentiële effect van prevalentie van specifieke PAO clades op de procesprestaties en de belangrijke rol van het EBPR proces in de verwijdering en terugwinning van nutriënten, was er een behoefte om de functionele verschillen tussen PAO clades te onderzoeken. Het doel van dit deel van de studie was om de functionele verschillen tussen PAO clades te onderzoeken waarbij voornamelijk het anaerobe metabolisme onderzocht is in relatie met de vorming van opslagpolymeren en het anoxische metabolisme.

In **Hoofdstuk 2** werd, met behulp van korte termijn experimenten, aangetoond dat er significante functionele verschillen bestaan tussen PAO I en II met betrekking tot het anaerobe metabolisme voor de opname van vluchtige organische vetzuren (VFA). Beide PAO clades waren in staat om hun metabolisme te veranderen van een gemengd polyfosfaat en glycogeen afhankelijke metabolisme naar een metabolisme dat volledig afhankelijk was van glycogeen. Hierbij waren de azijnzuur (HAc) opnamesnelheden van beide PAO clades aanzienlijk afgenomen. Dit effect was het grootst voor PAO I. Onder polyfosfaat gelimiteerde omstandigheden, was de HAc-opnamesnelheid van PAO II vier keer sneller dan die van PAO I, terwijl die van PAO I sneller was wanneer er voldoende polyfosfaat aanwezig was. Bovendien, onder ongelimiteerde polyfosfaat omstandigheden voor de anaerobe HAc-opname, in PAO II vond een gemengd metabolisme plaats dat voor het genereren van de benodigde energie deels afhankelijk was van glycogeen en deels van polyfosfaat. Dit is in contrast met PAO I die in veel grotere mate afhankelijk was van polyfosfaat. Deze bevindingen leveren een bijdrage aan het oplossen van de controverse rondom de verschillende stoïchiometrische en kinetische waarden die waargenomen zijn in EBPR systemen. Ook zijn de bevindingen relevant voor de ontwikkeling van operationele richtlijnen voor (gecombineerde chemische en) biologische fosfaatverwijderingsprocessen.

In **Hoofdstuk 3**, werd het effect van opslagpolymeren op het metabolisme van PAO II onderzocht in lange termijn experimenten. De verkregen resultaten werden vergeleken met een eerdere studie die, gebaseerd op de gerapporteerde stoichiometrie en kinetiek, vermoedelijk uitgevoerd was met een PAO I gedomineerde microbiële cultuur. De bevindingen die voort kwamen uit deze vergelijking, bevestigde de waarnemingen van de korte termijn experimenten met betrekking tot de functionele verschillen tussen PAO I en II. Ook biedt deze studie interessante inzichten in de rol van de opslag van polymeren op de regulering van het anaerobe HAc-opname metabolisme. Wanneer de influent P/C-verhouding verhoogd werd, nam het polyfosfaatgehalte van de biomassa toe terwijl het glycogeengehalte afnam. Bij verhoging van de polyfosfaatgehaltes, nam de fosfaat afgifte snelheid voor HAc-opname en voor maintenance toe. Tegelijkertijd met de toename in de fosfaatafgifte snelheden, namen ook de HAc-opname snelheden toe tot een optimale polyfosfaat/glycogeen verhouding van 0,3 P-mol/C-mol. Boven deze optimale verhouding, nam de de HAc-opname snelheid weer af. Aan de hand van de stoichiometrie van de anaerobe omzettingen is gebleken dat er een verandering in het metabolisme plaats vond. Het metabolisme veranderde van een glycogeen afhankelijk naar een polyfosfaat afhankelijk metabolisme wanneer het polyfosfaatgehalte van de biomassa toe nam. De veranderingen in de HAc-opname snelheden, suggereren dat bij lage polyfosfaatgehaltes de ATP productie de snelheidsbepalende stap is van het HAc-opname process, terwijl bij hoge polyfosfaatgehaltes (en dus lage glycogeen gehaltes) de NADH productiesnelheid de snelheidsbeperkende stap voor HAc-opname wordt. Elektronenmicroscopie toonde aan dat polyfosfaat opgeslagen wordt in de vorm van grote granules in de PAO cellen en daarom zou het kunnen zijn dat de polyfosfaatconsumptie snelheid beperkt wordt door het oppervlak van de granules. Hierdoor kan een afname van het polyfosfaatgehalte van de biomassa, de ATP productie snelheid beperken en kan hiermee de ATP productie door middel van glycogeen omzetting geactiveerd worden. De resultaten uit deze studie dragen bij aan een beter begrip van het metabolisme van PAO clades onder dynamische omstandigheden en bieden meer inzicht in de populatie dynamiek die waargenomen is in eerdere studies.

Ter bevestiging van de bevindingen in hoofdstuk 2 en 3, werd in **Hoofdstuk 4** onderzocht of bepaalde PAO clades in staat waren om in het systeem te groeien onder omstandigheden waarbij de influent fosfaatconcentraties net voldoende waren voor de assimilatie van biomassa (en dus beperkend voor de vorming van polyfosfaat). Voor dit doel werd een sequencing batch systeem geinoculeerd met actief slib en onder alternerende anaerobe en aerobe condities bedreven met synthetisch afvalwater. Na een periode van 16 slibverblijftijden werd er een gemengde PAO-GAO cultuur verrijkt waarvan de totale bacteriële populatie uit 49% PAO en 46% GAO bestond. De PAO waren nauw verwant aan de 'Candidatus Accumulibacter phosphatis' Clade II. Onder anaërobe omstandigheden, vond er in de gemengde PAO-GAO cultuur een typerend GAO metabolisme plaats waarbij alle energie voor HAc-opname werd geproduceerd door de omzetting van glycogeen. Deze studie bevestigt de bevindingen, in hoofdstuk 2 en 3, dat PAO een glycogeen afhankelijke metabolisme kan uitvoeren en dat PAO II een competitief voordeel heeft ten opzichte van PAO I onder fosfaat beperkende omstandigheden. Onder aërobe omstandigheden waren PAO II direct in staat om grote hoeveelheden fosfaat op te nemen wanneer extra fosfaat werd toegevoegd aan het influent. Vanuit een praktisch perspectief, toonde de studie aan dat in PAO II gedomineerde actief slib systemen, PAO II voor perioden van 16 slibverblijftijden in het actief slib kunnen blijven onder fosfaat beperkende condities bijvoorbeeld als gevolg van een ijzer overdosering. Wanneer de fosfaat concentraties weer toenemen zijn ze nog steeds in staat om fosfaat onder aëroobe omstandigheden op te nemen.

Hoofdstuk 5 is gericht op de denitrificatie routes van PAO. Verschillende literatuur studies suggereerden dat PAO I zowel nitriet als nitraat kan gebruiken als externe elektronenacceptor terwijl PAO II alleen in staat zou zijn om nitriet te gebruiken. De resultaten van die eerdere studies zijn tegenstrijdig en twijfelachtig aangezien er geen studies zijn uitgevoerd met EBPR culturen die sterk verrijkt zijn met specifieke PAO clades onder geschikte experimentele omstandigheden. Om deze tegenstrijdige resultaten te verklaren werden in hoofdstuk 5 de oxidatieve routes (zuurstof, nitriet en nitraat) van een PAO I cultuur onderzocht in combinatie met verschillende electron donoren (HAc en HPr). Deze combinaties werden onderzocht na een cultivatie periode in anaërobe/aërobe modus en vervolgens na een cultivatie periode in anaërobe/anoxische/aërobe modus. Na cultivatie in anaërobe/anoxische/aërobe modus was de verrijkte cultuur niet in staat om in de aanwezigheid van nitraat, fosfaat op te nemen, ondanks dat er lage denitrificatiesnelheden waargenomen werden. In aanwezigheid van zuurstof en nitriet, werd snelle fosfaat opname waargenomen. Het grote verschil in denitrificatie snelheden van nitriet en nitraat samen met de waarneming dat flankerende microbiele populaties nog steeds aanwezig waren in de sterk verrijkte biomassa cultuur, resulteerde in de hypothese dat de flankerende microbiele populaties in de biomassa verantwoordelijk waren voor de omzetting van nitraat naar nitriet, waarbij de koolstofbron voornamelijk verkregen werd van uitgestoten opgeloste microbiële producten. Deze hypothese werd verder ondersteund door een vergelijking van literatuurwaarden van studies met verrijkte PAO culturen, waarin de mate van PAO verrijking varieerde. Uit deze vergelijking van literatuurwaarden bleek dat de biomassa specifieke P-opnamesnelheid in aanwezigheid van nitraat toe neemt naarmate de fractie van flankerende microbiele populaties toeneemt en die van PAO dus afneemt. Verder bleek het dat onder de anaërobe/anoxische/aërobe cultivatie modus, de stoichiometrische verhouding van de P-verwijdering/N-verwijdering, met nitriet hoger was dan met nitraat. Dit stoichiometrische verschil suggereert dat PAO in het algemeen niet in staat is om nitraat te gebruiken als externe elektronenacceptor en voor zijn denitrificatie activiteit afhankelijk is van de gedeeltelijke denitrificatie-activiteit van andere organismen.

Dit eerste deel van het onderzoek heeft aangetoond dat er aanzienlijke functionele diversiteit bestaat in het metabolisme van PAO met betrekking tot het anaerobe metabolisme. Verder bleek het dat er met betrekking tot de denitrificatie routes tussen de accumulibacter clades PAO I en II er mogelijk geen functionele verschillen zijn. Deze verschillen in het anaerobe metabolisme dragen bij aan een beter begrip van de waarnemingen in eerdere studies. Met name de populatiedynamiek en vanuit een meer praktisch oogpunt ook voor de ontwikkeling van nutriënt terugwinning en/of gecombineerde chemische en biologische fosfaatverwijdering systemen. De bevindingen van de denitrificatie routes van PAO I geven meer inzicht in de rol van de PAO in gecombineerde nutriëntenverwijdering systemen. Ook bied het onderzoek meer inzicht in de rol van de opslag polymeren in de regulering van het anaerobe substraatopname metabolisme, wat ook leidt tot een beter begrip van EBPR processen in zuiveringsinstallaties onder dynamische omstandigheden.Hoewel deze studie inzicht geeft in de functionele diversiteit van PAO clades en hun metabolisme, is dit slechts het uitgangspunt van onderzoek gericht op de functionele diversiteit van PAO clades. Om toekomstig onderzoek naar de functionele diversiteit mogelijk te maken, moet een betrouwbare selectie of isolatie methoden voor PAO ontwikkeld worden om zeer verrijkte of zuivere culturen met specifieke PAO clades te verkrijgen.

Korte termijn invloed van zoutgehalte op het metabolisme van PAO en GAO (Hoofdstuk 6 en 7)

Zout afvalwater kan worden gegenereerd door de industrie, het infiltreren van zout water in de riolering of als zout water direct gebruikt wordt als alternatieve waterbron voor doeleinden die geen drinkwater vereisen zoals het doorspoelen van toiletten. Om het milieu te beschermen tegen ernstige milieuproblemen zoals hypoxie en eutrofiëring, moeten de nutriënten (C, N, P) verwijderd worden uit zoute afvalwater stromen voor de lozing op de ontvangende wateren. Echter, het zoutgehalte kan negatieve invloed hebben op de micro-organismen die verantwoordelijk zijn voor de verwijdering van nutriënten in systemen voor biologische verwijdering van nutriënten. Om deze reden werd in deze studie onderzocht wat het effect van het zoutgehalte is op het metabolisme van de microbiële populaties die heersen in EBPR systemen (PAO en GAO).

In **Hoofdstuk 6** werden de korte termijn effecten van verhoogde zoutgehaltes op het anaërobe metabolisme van PAO en GAO bestudeerd. Er werd aangetoond dat een verhoogd zoutgehalte een negatief effect had op zowel PAO als GAO, waarbij PAO het meest gevoelige organisme was. Dit negatieve effect bestond uit de remming van HAc opname snelheid bij toenemende zoutgehaltes terwijl de maintenance activiteit toenam (tot een zoutgehalte van 4%) voor zowel de PAO als de GAO. Verhoogde zoutgehaltes leken ook een verandering in het metabolisme van PAO teweeg te brengen, namelijk de verschuiving van poly-P consumptie naar glycogeen consumptie voor de energie productie voor de opname van HAc en maintenance. De stoichiometrische verhoudingen van de anaerobe HAc-opname omzettingen van GAO werden niet beïnvloed. Verder werd er een gestructureerd model ontwikkeld, dat met succes de effecten van verhoogde zoutgehaltes op de verschillende metabolische processen van PAO en GAO kon beschrijven.

In **Hoofdstuk 7**, zijn de korte-termijn effecten van verhoogde zoutgehaltes op het aerobe metabolisme van PAO bestudeerd. Door dit onderzoek werd duidelijk dat het aerobe metabolisme zeer gevoelig was voor lage zoutgehaltes. Een verhoging in het zoutgehalte van 0,02 naar 0,18% leidde tot een daling van de specifieke O_2, PO_4 (polyfosfaat productie) en NH_4 (biomassa productie) opnamesnelheden van 25%, 46% en 63%, respectievelijk. Bij een zoutgehalte van 0,35% of hoger, werden de PO_4, NH_4 opname en glycogeen productie volledig geinhibeerd. De groei van biomassa was het meest geinhibeerde proces, gevolgd door poly-P productie en glycogeen productie. Verder nam de aërobe energie productie voor maintenance toe tot een drempel zoutgehalte van 2%, waarboven de energie productie voor maintenance snel afnam. Om extra energie te genereren voor de toenemende energiebehoefte voor maintenance, werd er fosfaat afgegeven bij zoutgehaltes van meer dan 0,35%. Deze aërobe fosfaatafgifte volgde een soortgelijke trend als het maintenance zuurstofverbruik. Het inhibitie model dat in deze studie ontwikkeld is, kon de waargenomen effecten van verhoogde zougehaltes op de verschillende metabolische processen in deze studie beschrijven.

Het tweede deel van dit onderzoek heeft aangetoond dat het EBPR proces, met name de aërobe fase zeer gevoelig kan zijn voor verhoogde zoutgehaltes. De bevindingen suggereren dat een verhoging van de zout concentraties boven de 0.35%, ernstige verstoringen van het EBPR process kan veroorzaken. Een dergelijke verhoging van de zout concentraties kom overeen met een lozing van 5% zeewater (met 3,4% zoutgehalte) of 15% brak water (met 1,2% zoutgehalte), door toiletspoeling met zeewater, industriële lozingen of infiltratie van zoutwater in de riolering. Dit geeft aan dat het EBPR process mogelijk niet toepasbaar is voor de behandeling van zout afvalwater en dat het zoutgehalte een relevante inhibitie factor kan zijn in actief slib systemen die conventioneel huishoudelijk afvalwater behandelen. De gepresenteerde gegevens zijn echter gebaseerd op de korte termijn experimenten met natriumchloride. Om de organismen de gelegenheid te geven om te wennen aan hogere zoutgehaltes of om het systeem de mogelijkheid te geven om meer zouttolerante PAO clades te selecteren, dienen toekomstige studies gericht te worden op de lange termijn effecten van verhoogde zoutgehaltes op EBPR culturen, waarbij er rekening gehouden moet worden met de samenstelling van het zout en van het synthetische afvalwater.

1

General introduction

1.1 Introduction

This PhD study focuses on the enhanced biological phosphorus removal (EBPR) process in activated sludge systems. The first part of the thesis is oriented to the functional diversity among polyphosphate-accumulating organisms clades, while the second part of the thesis addresses the salinity effects on the metabolism of polyphosphate-accumulating organisms (PAO) and glycogen-accumulating organisms (GAO). This chapter provides background information of the EBPR process and introduces the relevance of the presented PhD research.

1.2 Enhanced biological phosphate removal (EBPR)

Enhanced biological phosphorus removal is a microbial process for removal of excessive amounts of phosphate from wastewater through storage of intracellular poly-P by polyphosphate-accumulating organisms and excess sludge wasting (Figure 1.1). Due to its high efficiency and cost-effectiveness, the process is widely implemented in biological wastewater treatment systems. The PAO are able to take up phosphate from the liquid phase and store it as intracellular polyphosphate, leading to P-removal from the bulk liquid via PAO cell removal through the wastage of activated sludge. Polyphosphate-accumulating organisms have the potential to store 0.38 mgP/mgTSS (Wentzel et al., 1987; Wentzel et al., 1988; Schuler and Jenkins et al., 2003), versus 0.023 mgP/mgVSS for ordinary heterotrophic organisms (Metcalf and Eddy, 2003). In the EBPR process, the development of polyphosphate-accumulating organisms (PAO) in the sludge is favored by cycling it through anaerobic and aerobic/anoxic stages (Barnard *et al.*, 1975). Since the first indications of a microbial process for EBPR (Srinath et al., 1959; Levin and Shapiro, 1965), several different process configurations have been developed. Figure 1.1a displays the simplest configuration, developped for simultaneous BOD and P removal making use of an anaerobic and oxic stage. For simultaneous BOD, N and P removal, an additional anoxic stage is required as shown in Figure 1.1b. More advanced configurations have been developped to optimise the combined nitrogen and phosphate removal process, and to combine biological and chemical phosphate removal from wastewaters that have unfavourable influent compositions for complete biological phosphate removal. Dependent on the wastewater composition, climate, process configurations and others, the operational conditions in full scale wastewater treatment systems vary over a wide range. For full scale wastewater treament systems, that include an EBPR process, solid retention times can vary from 3 days up to 30 days in activated sludge systems or even longer in granular sludge systems. In a survey of 7 treatment plants in the Netherlands with different type of configurations, the nomical total hydraulic retention times varied from 10 hours to 52 hours (Lopez-Vazquez et al., 2008) with the anaerobic, anoxic and aerobic hydraulic retention times ranging from 0.5-12, 0-12 and 7-46 hours, respectively. Typical sludge contact times in the anaerobic stages of wastewater treatment plants in the Netherlands can vary from 0.2-3 hours (Janssen et al., 2002).

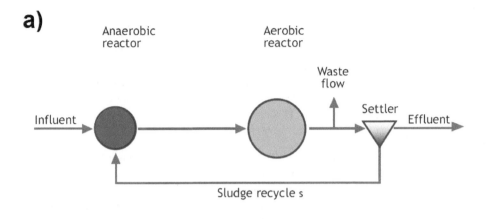

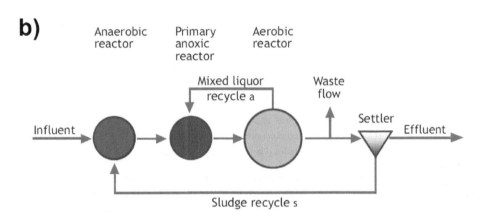

Figure 1.1 Schematic representations of simple process configurations for BOD and P removal (anaerobic/oxic system)and combined BOD, N and P removal (anaerobic/anoxic/oxic system). Figure adopted from Henze et al. (2008).

1.2.1 PAO metabolism

In the anaerobic phase, unlike most of other microorganisms, PAO can take up volatile fatty acids (VFA), such as acetate (HAc) and propionate (HPr), and store them intracellularly as carbon polymers, namely poly-β-hydroxyalkanoates (PHA) (Wentzel *et al.*, 1985, Comeau *et al.*, 1986, Mino *et al.*, 1987). Dependent on the type of carbon source (and its degree of reduction) fed to the biomass, different type of PHA can be formed, such as poly-β-hydroxybutyrate (PHB), poly-β-hydroxyvalerate (PHV), poly-β-hydroxy-2-methylvalerate (PH2MV) and poly-β-hydroxy-2-methylbutyrate (PH2MB). This unique metabolism gives PAO a competitive advantage over organisms that do not have the ability to store organic matter in the anaerobic phase. For the uptake of VFA and their transformation into PHA, energy and reducing power are required. The energy is mainly generated by the cleavage of intracellular polyphosphate (poly-P) and subsequent release of phosphate over the cell membrane. For the production of reducing power and additional energy, different pathways are proposed, such as the glycolysis of intracellularly stored glycogen (Mino *et al.*, 1987; Smolders *et al.*, 1994), the operation of the TCA cycle (Comeau *et al.*, 1986; Wentzel *et*

3

al., 1986), the combination of both the glycolysis and TCA cycle (Pereira *et al.*, 1996), among others. Moreover, there is contradicting evidence regarding the glycolysis. Both the Entner-Doudoroff (ED) pathway (Maurer *et al.*, 1997; Hesselman *et al.*, 2000) and the Embden-Meyerhoff-Parnas (EM) pathway (Martin *et al.*, 2006) have been proposed. In the aerobic (or anoxic) phase, PAO utilize the anaerobically stored PHA as carbon and energy source for maintenance, growth, glycogen formation and poly-P formation. For poly-P formation, P is taken up from the bulk liquid and synthesized into poly-P. Due to PAO biomass growth, the aerobic P-uptake is higher than the anaerobic P-release resulting in a net P-removal from the liquid phase (Mino *et al.*, 1998).

1.2.2 Identity of PAO

Although numerous studies have focused on EBPR and the microorganisms responsible for EBPR, PAO bacteria have not been isolated yet. By linking microbial community composition with EBPR performance, bacteria from the subclass 2 *Betaproteobacteria*, closely related to *Rhodocyclus* were identified as the organisms responsible for the PAO metabolism observed in laboratory EBPR systems (Bond *et al.* 1995, Bond et al., 1999; Hesselman et al., 1999; Crocetti et al., 2000). These organisms were named "*Candidatus* Accumulibacter phosphatis" In several studies it was confirmed that "*Candidatus* Accumulibacter phosphatis", observed in laboratory studies, was also significantly present in activated sludge of several wastewater treatment plants with various process configurations across four continents (Zilles et al., 2002; Saunders et al., 2003; Kong et al., 2004; Gu et al., 2005; He et al., 2005; Wong et al., 2005). In addition, Saunders et al. (2003) observed a good correlation between the EBPR performance of six full scale treatment plants in Australia and the percentage of Accumulibacter present in the sludge, supporting that these are the organisms responsible for EBPR in laboratory studies as well as in full scale treatment plants. Using both the 16SrRNA gene and the poly-phosphate kinase gene (ppk1) as a genetic marker, it was revealed that *Accumulibacter* was organized into two main clades, *Candidatus* Accumulibacter phosphatis clade I (PAO I) and *Candidatus* Accumulibacter phosphatis clade II (PAO II). Both clades I and II comprised of several distinct sub-clades (He *et al.*, 2007; Peterson *et al.*, 2008).

1.2.3 Functional differences of PAO clades affecting the EBPR performance

Although the genetic diversity of Accumulibacter clades has been observed in several studies, the relation with and existence of functional diversity of Accumulibacter clades is unclear. There are some indications that functional differences exist that may affect the process performance. For instance, Carvalho et al. (2007) and Flowers *et al.* (2009) reported that clade IA was able to couple nitrate reduction with phosphorus uptake, but clade IIA was not. Slater *et al.* (2010) monitored associations between clade-level variations, overall community structure and ecosystem function in EBPR systems using terminal-restriction fragment length polymorphism (T-RFLP). A shift in the ratio of the dominant *Accumulibacter* clades was detected, with type IA associated with good EBPR performance and type IIC associated with poor EBPR performance. However, only a limited number of EBPR studies reported their findings in association with specific PAO clades, as techniques for identification of different clades have been developed relatively recent and operational conditions for selection of specific PAO clades have not been developed yet. Possible existence of functional differences among PAO clades implies that the findings in laboratory studies may depend partially on the type of PAO clades, enriched in those studies and could help to explain many contradictions regarding the metabolic pathways and physiological properties of PAO. Furthermore, for the operation of EBPR systems, it is relevant to understand the functional differences between PAO clades because the operational conditions may select specific PAO clades with specific metabolic abilities such as the ability to proliferate in systems during periods of low influent phosphate/BOD ratios (industrial wastewaters), the use of nitrate as external electron acceptor (denitrifying EBPR systems) or the tolerance to stress conditions such as low pH or high salinity. The lack of knowledge on the functional differences among PAO clades and lack of knowledge on the microbial community composition of enrichment cultures in past studies hampers the interpretation of those studies and the translation of the findings to practical applications. Therefore it is necessary to obtain a better understanding about functional differences among PAO clades and report in future studies the composition of the microbial communities.

1.2.4 Glycogen accumulating organisms (GAO)

Although PAO have strong competitive advantage over ordinary heterotrophic organisms (OHO) when sludge is cycled through anaerobic and aerobic stages, another group of organisms co-exists, so-called glycogen accumulating organisms (GAO), that have the ability to proliferate under similar conditions as PAO. In order to do so, these organisms use a GAO metabolism which is similar to the PAO metabolism, except that poly-P is not involved in it, and therefore GAO do not contribute to excessive P-removal (Mino *et al.*, 1987).

1.2.5 GAO metabolism

In the anaerobic phase, like with PAO, the GAO take up VFA and convert it to PHA. However, the energy that is needed for this process is, unlike with PAO, fully generated by the conversion of glycogen into PHA. Therefore, GAO utilize glycogen to a much larger extent as compared to PAO. For the generation of reducing power, glycolysis of intracellularly stored glycogen is considered as the main pathway (Filipe et al., 2001b; Zeng et al., 2002), while partial involvement of the TCA cycle has also been suggested (Saunders et al., 2007; Lemos et al., 2007; Burrow et al., 2008). One consequence of using glycogen for the production of energy is the higher production of reduction equivalents. To maintain the intracellular redox balance, GAO produce more reduced PHA forms, such as PHV (Liu *et al.*, 1994; Satoh *et al.*, 1994; Filipe *et al.*, 2001a; Zeng *et al.*, 2003a) while PAO produce mainly PHB when fed with acetate, with little PHV production (Satoh *et al.*, 1992; Smolders *et al.*, 1994; Mino *et al.*, 1998). Under aerobic conditions, GAO use the stored PHA as carbon and energy source similar to the PAO metabolism for maintenance, growth and glycogen formation, with the only difference that there is no intracellular poly-P formation (Filipe *et al*, 2001b; Zeng *et al.*, 2003a). On the basis of the differences in the carbon and phosphorus conversions between the PAO and GAO metabolism, the stoichiometric parameters of the anaerobic carbon and phosphorus conversions were often used as indicators for the fractions of PAO and GAO presence in the total bacterial population (Oehmen et al., 2007; Lopez-Vazquez et al., 2007).

1.2.6 Identity of GAO

Two different groups of organisms were found that are able to perform the GAO metabolism. Similar to the PAO, these organisms have not been isolated yet. In deteriorated laboratory EBPR systems, organisms belonging to the γ-proteobacteria were the first to be identified as organisms performing a GAO metabolism (Nielsen et al., 1999; Crocetti et al., 2002). These organisms were named "*Candidatus* Competibacter phosphatis". A phylogenetic analysis of 16srRNA gene sequences obtained from several studies revealed that these bacteria formed a novel cluster in the γ-proteobacteria with seven subgroups, showing the existence of a wide genetic diversity among GAO (Kong et al, 2001). After the development of fluorescence *in situ* hybridization (FISH) probes, additional studies demonstrated that these "*Candidatus* Competibacter phosphatis" were found dominantly in many acetate fed laboratory systems (Crocetti et al., 2002; Kong et al., 2002b; Zeng et al., 2003b; Oehmen et al.,2004) and were also significantly present in the sludge of full-scale plants (Crocetti et al., 2002; Kong et al., 2002b;Saunders et al., 2003; Gu et al., 2005; Wong et al., 2005; Kong et al., 2006). In a later stage, another group of organisms, belonging to the α-proteobacteria, was identified that could perform a GAO metabolism in laboratory systems with poor EBPR activity. This group was closely related to Defluviicoccus vanus (Wong et al., 2004; Meyer et al., 2006). Phylogenetic analysis showed that the defluviicoccus related GAO formed a monophyletic group with two distinct clusters. While α-proteobacteria GAO have been found in many laboratory studies, the organisms were only detected in low abundance in full scale wastewater treatment plants (Burow et al., 2007; Lopez-Vazquez et al., 2008).

Recent studies suggest that in addition to Competibacter and Defluviicoccus, PAO (or certain PAO clades) are also able to perform a GAO metabolism in short-term batch tests under poly-P limiting conditions (Zhou et al., 2008; Acevedo et al., 2012). This implies that the stoichiometry of the carbon and phosphorus conversions may no longer be a good parameter to determine the fractions of PAO and GAO clades and that verification by staining techniques or molecular techniques is required to verify quantitatively and qualitatively which populations are present in the systems.

1.3 Factors affecting the EBPR process performance

The performance of EBPR systems is dependent on the prevalence of PAO in the sludge. Many studies were focused on the conditions that give PAO a competitive advantage over ordinary heterothophic organisms (OHO) and GAO to support the prevalence of PAO and thereby improve the process performance.

1.3.1 Factors affecting the competition between PAO and OHO

Considering the unique metabolism of PAO, it is important to expose the sludge in a strict anaerobic zone to wastewater containing VFA, followed by an aerobic phase. This can give PAO a competitive advantage over OHO and can lead to the prevalence of PAO. If nitrate and/or oxygen are introduced into the anaerobic tank, the biological phosphate removal process deteriorates (Barnard et al., 1976; Hascoet and Florentz, 1985), due to a direct substrate competition between OHO and PAO. Other factors such as the presence of magnesium (Mg) and potassium (K), which serve as counterions in poly-P, seemed to be relevant for the development of PAO biomass (Brdjanovic et al., 1998).

1.3.2 Factors affecting the competition between PAO and GAO

Since the discovery of PAO (Mino *et al.*, 1987), many studies focused on the environmental factors affecting the competition between PAO and GAO. In a extensive review, Oehmen *et al.* (2007) pointed out that, temperature, pH and carbon source were important factors in the PAO-GAO competition and suggested that sludge age (SRT), dissolved oxygen (DO) concentrations and accumulation of nitrite could affect the PAO-GAO competition. Lopez-Vazquez *et al.* (2009) confirmed that the competition between PAO and GAO is affected by temperature, carbon source and pH. According to their research, GAO are only able to compete with PAO for substrate at temperatures higher than 20°C. At 20°C, propionate and acetate should be present to favor PAO growth irrespective of the pH value. In the case that only acetate or propionate is used as carbon source, the pH should be above 7.5 to facilitate PAO growth. In another study, it was demonstrated that free nitrous acid (which is one of the forms of nitrite) can negatively affect the metabolism of PAO leading to the deterioration of the EBPR process (Pijuan *et al.*, 2010). Moreover, since GAO appear to be less sensitive to the presence of free nitrous acid, its accumulation can favor GAO over PAO (Pijuan *et al.*, 2010; Ye *et al.*, 2010; Zhou *et al.*, 2010), leading to the deterioration of the EBPR process. Although the effect of several operational conditions (temperature, pH, carbon source, SRT, DO and free nitrous acid concentration) was studied in detail on PAO and their competition with GAO, the effect of another important factor - salinity - has not been addressed yet.

1.4 Salinity as a factor affecting microorganisms

1.4.1 Generation of saline wastewater

In the past century, the natural water cycle has been affected drastically by human intervention. Due to population growth, there is an increasing water demand for different types of activities, like industrial, agricultural and residential. Almost all activities lead to the generation of wastewater. Dependent on the water supply options, the type of activities and wastewater collection options, those wastewaters may contain higher salinity levels than the usual level. In case of residential needs, water is used for human consumption, washing, as a means to convey waste (feces and urine) through a sewerage system to a collection point for further treatment and/or disposal, cooling and other purposes. When traditional water supply solutions (including rainwater harvesting, water diversions, water storage in reservoirs and water transport from far away) become insufficient and non-feasible to match the water demand, the use of saline (sea and brackish) water as secondary quality water for flushing toilets or cooling can be a promising, cost-effective and environmentally-friendly alternative to alleviate fresh water stress in urban areas located in coastal zones. The energy requirements for the production of secondary quality water from seawater are in the range of 0.013-0.025 (kWh/m^3), compared to 0.05, 0.2-1.0 and 2.5-6.0 (kWh/m^3) for conventional, water reclamation and seawater desalination systems respectively (Leung et al., 2012). Up to 30% of the fresh water used by households can be replaced by saline water for toilet flushing (Jiang, 2004; Mayer et al., 1999; Foekema et al., 2008). In addition, salinity serves as a tracer that eliminates the risk of cross connection between potable and non-potable water supply systems. The benefits of using seawater as secondary quality water have been successfully demonstrated in Hong Kong (Tang et al., 2006; Leung et al., 2012), where the use of seawater for flushing toilets was introduced about 50 years ago. In addition to saline water usage for toilet flushing, there are industries that use saline water or generate saline water in certain processing steps, which leads to the generation of saline wastewaters (Gonzalez et al., 1983; Orhon et al., 1999; Fahim et al., 2000; Lefebvre et al., 2006). Furthermore, sprinkling of salt on the roads to prevent ice formation and thereby mitigate slip hazards in winter time, will lead to an increase of salinity in storm water, which is often collected in combined sewerage systems, where it gets mixed with domestic wastewater during its transportation. Finally, the intrusion of saline water, such as brackish ground water or seawater during high tides, into the sewerage systems leads to increased salinity levels in the wastewaters.

1.4.2 Effect of salinity on microorganisms

When micro-organisms are exposed to salinity, there are several physical/chemical stress factors they have to deal with such as osmotic pressure, ionic strength (Brown 1990, Galinski and Truper, 1994; Sleator and Hill, 2001; Measures, 1975; Roesler and Muller, 2001) and large concentration gradients of different types of ions (Castle et al., 1986).

In order to overcome the osmotic stress, osmotic equilibrium is required (Truper and Galinski, 1986; Tindall, 1988; Larsen, 1986; Gilmour, 1990). There exist two main osmo-adaptation strategies for obtaining the equilibrium: the 'salt in cytoplasm' type and the 'compatible solute' type (Galinski and Truper, 1994; Sleator and Hill, 2001; Measures, 1975; Roesler and Muller, 2001). With the 'salt in cytoplasm' type the cells maintain osmotic equilibrium at high salinity by importing large amounts of potassium chloride KCl into the cytoplasm. In order to survive the high internal salt concentration, extensive structural adaptations are required due to the ionic strength. Organisms employing the 'salt in cytoplasm' type of osmoadaptation, can only adapt to changes in the salinity concentrations within a narrow range of salinity levels (Galinski and Truper, 1994). Representatives of the group of organisms that use the 'salt in cytoplasm' strategy are the halobacteria (Archea), eubacterial fermenting and/or acetogenic anaerobes (Zhilina and Zavarzin, 1990; Oren, 1991) and sulfate reducers (Caumette et al., 1991;

Ollivier *et al.*, 1991). With the 'compatible solute' type osmoadaptation, the cells maintain the osmotic equilibrium by production or import of small organic molecules called 'compatible solutes' or 'osmolytes' and its subsequent accumulation. Compatible solutes are defined as: small organic solutes that are highly soluble and carry no net charge at physiological pH (Galinski, 1995) and do not interact with proteins. Therefore they can be accumulated to high intracellular concentrations (>1 mol kg^{-1} water (Brown, 1976; Galinski and Truper, 1994)) without severely affecting vital cellular processes (Record *et al.*, 1998; Yancey, 1994). When the 'compatible solute' type of adaptation is applied, a 'normal' salt-sensitive enzymatic machinery is preserved (Galinski and Truper, 1994) and extensive structural adaptations of the cells interior are not needed to survive salinity. The 'compatible solute' osmoadaptation strategy allows adaptation to salinity over a wide range of salinities, as the organisms can regulate the accumulation of intracellular compatible solute concentrations. . The 'compatible solute' type of osmoadaptation is wide spread and observed in all three domains of life (Bohnert, 1995; Kempf and Bremer, 1998; Roberts, 2000). Therefore it is more likely to be adopted by the organisms subjected in this study. When cells are exposed to a saline environment, a bi-phasic response occurs (Sleator and Hill, 2001). As a primary response mechanism, microorganisms increase the level of potassium. As a secondary response mechanism, microorganisms increase dramatically their cytoplasmic concentration of compatible solutes in response to increased intracellular potassium concentrations (Yancey *et al.*, 1982; Brown, 1976).

To deal with the ionic strength, micro-organisms need to adapt their enzymes and membrane lipids. Micro-organisms that use the 'compatible solute' type of osmoadaptation, only have their exterior exposed to the high ionic strength and need to adapt their membrane lipids (Galinski and Truper, 1994). When exposed to salinity, a change in the composition of the membrane lipids is often observed. The anionic lipids (phosphatidylglycerol and/or glycolipids) seem to increase relative to zwitterionic lipids. This has a pronounced effect on lipid phase behaviour (Russel, 1989; Sutton *et al.*, 1991). In the case that the micro-organism posses the 'salt in cytoplasm' type of osmoadaptation mechanism, the internal salt concentration is high, exposing the entire content of the cell to high ionic strength. These microorganisms need their entire cell interior adapted to ionic strength. In extreme halophilic microorganisms, in which the 'salt in cytoplasm' mechanism was found, the enzymes are extensively adapted. In general more acidic amino residues are observed in the enzymes of these organisms (Lanyi, 1974). The acidic groups are thought to enhance the attraction of a hydration shell in an environment with low water activity. Furthermore, the ions are thought to affect the tertiary and quaternary structure of enzymes (Sleator and Hill, 2001).

Finally, high concentration gradients of different ions may lead to leakage of ions over the membrane of the cells. Once ions enter the cells, they will need to be expelled from the intracellular environment as they may interfere with metabolic processes in different ways. Therefore the cells will have to export the ions against the concentrations gradient on the expense of energy (Castle et al., 1986), which could lead to futile cycles.

Overall, when the salinity concentration increases, bacteria need to balance the osmotic pressure, may need to change the enzymes and membrane lipids and maintain concentration gradients. These processes will likely require adaptation time and possibly selection of specific salt-tolerant PAO clades. Therefore the momentary salt effects on the metabolism of PAO and GAO may be different from the long-term salt effects. Whether the 'compatible solute' type or the 'salt in cytoplasm' type of osmo-adaptation is used to reach osmotic equilibrium, both mechanisms require energy and may lead to a decrease in the active biomass yield (Oren 1999). Regarding the 'compatible solute' type of osmoadaptation, the organisms may not be able to produce compatible solutes from their main carbon source, but may still have the ability to produce them from precursors supplied in the medium or just accumulate them by importing them from the extracellular environment. Therefore the composition of the medium or wastewater may be an important factor affecting the salinity tolerance of PAO and GAO. In addition, potassium serves as a signaling molecule to trigger osmo-adaptation processes (Yancey *et al.*, 1982; Brown, 1976) and therefore, the potassium over sodium ratio of the saline water may affect the ability of the organisms to sense the presence of salinity and acquire tolerance to salinity. When enzymes are affected by the ionic strength and/or lower water activity, the enzymes need to be adapted and/or all kinetic rates as well as stoichiometric parameters may be affected. Maintaining concentration gradients of different type of ions will probably also require energy, increasing the maintenance energy requirements and decreasing the active biomass yield of PAO and GAO.

1.4.3 Impact of salinity on biological nutrient removal

When saline wastewaters need to be treated by biological processes, the salinity may affect the microorganisms responsible for different nutrient removal processes. An evaluation of previous studies (Uygur et al., 2006, Moussa et al., 2006, Mariangel et al., 2008; van den Brand, 2014) revealed that on the basis of long-term experiments the activity of the microbial populations responsible for COD removal, nitrification and denitrification are all affected by the elevated salt content in the wastewater (Figure 1.2). However, the carbon (COD) removal, reflected on the activity of ordinary heterotrophic organisms (OHO) practically does not show any major deleterious effect up to 6% salinity (Uygur et al., 2006). The carbon COD conversion to sulfide COD, reflected on the activity of sulphate reducing bacteria (SRB) was inhibited by 50% when the salinity concentrations increased from 0.7% to 2.1% salinity. The nitrification process, the most sensitive step in conventional N removal (exemplified by ammonium oxidizing bacteria: AOB and nitrite oxidizing bacteria: NOB) was only reduced by 20-30% at 1% salinity, but it becomes more pronounced at 2% salinity (reaching around 50-60% reduction in comparison to the starting situation with almost zero salinity) (Moussa et al., 2006). According to the results presented here, the AOB are more affected by salinity than NOB. However, on the contrary, in several studies it has been reported that the NOB were more affected then the AOB, leading to the accumulation of nitrite (NO_2^-) in the system (Vredenbregt *et al.*, 1997; Dincer and Kargi, 1999; Intrasungkha *et al.*, 1999; Wu *et al.*, 2008; Cui *et al.*, 2009). Denitrification, which is usually considered to be performed in activated sludge systems by denitrifying ordinary heterotrophic organisms (OHO), is less than 20% inhibited at 1% salinity and only about 40% inhibited at 4% salinity (Mariangel et al., 2008). Only a few studies have been conducted to assess the salinity effects on the EBPR process but the outcomes from these studies are inconclusive and inconsistent, which is discussed in more detail in the next paragraph (1.4.4).

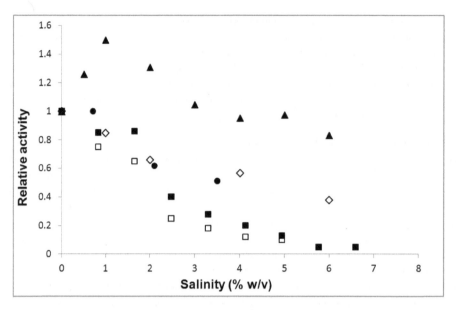

Figure 1.2 Impact of salinity on the relative specific nutrient removal activities of the microbial populations responsible for BOD and N removal in conventional wastewater treatment systems. (▲) OHO (Uygur et al., 2006), (◇) OHO (Mariangel et al., 2008), (□) AOB (Moussa et al., 2006), (■) NOB (Moussa et al., 2006), (●) SRB (van den Brand, 2014).

1.4.4 Impact of salinity on EBPR

Similar to the salinity effects on the BOD removal, nitrification and denitrification processes, the EBPR process may also be affected. However, only a few studies (Panswad and Anan, 1999; Intrasugkha et al., 1999; Uygur and Kargi, 2004; Kargi and Uygur, 2005; Kargi and Uygur, 2006; Hong et al., 2007; Wu et al., 2008; Cui et al., 2009) have focused on the effect of the salinity on the EBPR process and the results from these studies were inconsistent and inconclusive because (i) often suboptimal cultivation conditions were used for the growth of PAO, (ii) the impact of salinity on the biological carbon or nitrogen removal processes side effects may have affected PAO as the studies were not conducted with highly enriched PAO and GAO cultures, (iii) parameters relevant to assess the salinity effects on PAO and GAO were often not measured, and (iv) the specific PAO clades prevailing in the sludge were not reported. In several studies the cultivation conditions were suboptimal for PAO, even at 0% salinity, and therefore the EBPR performance was not satisfactory. In the studies from Uygur and Kargi (2004), Kargi and Uygur (2005) and Uygur (2006), there were indications that a significant part of the biological phosphorus removal (BPR) took place by assimilation of phosphorus for biosynthesis and not by enhanced uptake of phosphate (EBPR) for energy storage by PAO. Panswad and Anan (1999) achieved a phosphorus removal efficiency of only 48% at 0% salt concentration. Likely, a competition between GAO and PAO may have taken place. The temperature of 27-33°C applied in that study could have been the key factor. In the study of Intrasungkha et al. (1999), initially a competition between PAO and OHO took place due to the presence of nitrate and nitrite in the anaerobic phase. Extra addition of acetate to the influent to alleviate the PAO-OHO competition for volatile fatty acids (VFA) led to full phosphorus removal. In other studies, no VFA were added to the influent (Hong et al., 2007; Wu et al., 2008). In several studies (Abughararah and Sherrard, 1993; Panswad and Anan, 1999; Uygur and Kargi, 2004; Kargi and Uygur, 2005; Kargi and Uygur, 2006; Hong et al., 2007) the potential presence and accumulation of nitrite was not determined. Nitrite accumulation is often observed in nitrification processes operated at (high) saline conditions since nitrite oxidizing bacteria (NOB) are more sensitive to salinity than ammonium oxidizing organisms (AOB) (Dincer and Kargi, 1999; Vredenbregt *et al.*, 1997; Intrasungkha *et al.*, 1999; Wu *et al.*, 2008; Cui *et al.*, 2009). In a study with granular sludge, it was demonstrated that elevated salinity concentrations (21g/L NaCl) led to a cascade inhibition effect, where the deterioration of nitrite oxidation resulted in nitrite accumulation which in turn severely affected EBPR (Pronk et al., 2014). In most of the studies, data that could be useful to undertake a better assessment of the impact of salinity on the metabolism of PAO (like the anaerobic P-release and anaerobic C-uptake rates) are not reported (Panswad and Anan, 1999; Intrasugkha *et al.*, 1999; Uygur and Kargi, 2004; Kargi and Uygur, 2005; Uygur et al., 2006; Hong *et al.*, 2007). Therefore, it is hard to assess and evaluate, even at 0% salinity, to what extent EBPR took place and if it took place at all. Finally, none of these studies have reported which specific PAO clades were prevailing in the systems and considering potential differences in the metabolic response of PAO clades to changes in the environmental conditions, the outcomes of saline studies may be dependent on the specific PAO clades. Overall, based on the results regarding the impact of salinity on the EBPR process and the potential factors that might have affected the process in past studies, it was still unclear how salinity affects the EBPR process.

1.5 Scope of thesis

The first part of this thesis focuses on the functional diversity among PAO clades whereas the second part is directed to the salinity effects on the metabolism of PAO and GAO.

1.5.1 Functional diversity of PAO clades

Although genetic diversity among PAO clades (PAO I and PAO II) has been observed in past studies, PAO were often considered to behave functionally the same and therefore differentiation of the PAO clades in laboratory studies and modeling approaches was not considered useful. Recent studies have suggested that PAO clades may be functionally different regarding aspects such as the glycolytic pathways (ED or EMP), the ability to take up substrate under poly-P depleted conditions, the ability to use nitrate as external electron acceptor, and tolerance to stress conditions. Clarification of potential existence of such functional differences is important, as it suggests that the type of PAO clades may affect the process performance and therefore, it is necessary to clarify the main differences among PAO clades. Considering the significant role of the EBPR process in nutrient removal and recovery processes and the potential effect of PAO clades prevalence on the process performance, there is a need to verify the existence and type of functional differences among PAO clades. The objective of this part of the study was to assess the functional difference among PAO clades regarding the anaerobic metabolism in relation to their storage polymers and regarding the denitrification pathways. The anaerobic metabolism was selected to study the functional diversity, because in the anaerobic phase of the process different PAO clades and GAO populations compete for substrate and therefore the anaerobic metabolism plays a major role in population dynamics. Understanding the functional difference between PAO clades anaerobic metabolisms, may therefore help to link operational conditions with population dynamics and the process performance associated with specific populations. Considering the observations in recent studies that PAO can shift their metabolism from a poly-P dependent metabolism to a glycogen dependent metabolism, suggests that it is necessary to compare the differences in the anaerobic metabolism of PAO clades over a range of different poly-P contents to be able to distinguish between differences due to the poly-P content and potential intrinsic metabolic differences of PAO I and PAO II. The specific research questions of this part of the study are as follows:

1) What are the functional differences in the anaerobic physiology of PAO I and PAO II over a range of different poly-P contents during short-term experiments?
2) What is the effect of the biomass internal storage polymers (like poly-P and glycogen) on the metabolism of PAO I and PAO II during long-term experiments?

3) Do PAO I have the exclusive metabolic ability to use nitrate as external electron acceptor under anoxic conditions?

To fulfill the objective of this part of the study, an attempt was made to enrich specific PAO clades by running several bioreactor systems in parallel for a long period of time with different operational conditions, i.e. temperature, pH, VFA composition and influent P/C ratio. Once specific PAO clades were enriched and analysed by FISH and denaturing gradient gel electrophoresis (DGGE) analysis, their anaerobic metabolism was investigated in short-term and long-term experiments with different poly-P/VSS. In addition, it was verified if PAO I has, unlike PAO II, the ability to use nitrate as external electron acceptor under anoxic conditions. This study provides better insight in (i) the physiology of PAO clades (ii) population dynamics of different PAO clades and (iii) from a practical perspective, this may support the development of better metabolic models of the microbial populations prevailing in enhanced biological phosphorus removal systems.

1.5.2 Salinity effects on the metabolism of PAO and GAO

Due to different human activities and influences on the water cycle, wastewaters are generated that have increased salinity levels. In view of the ever growing global water demand, the direct use of saline water as secondary quality water for non-potable purposes seems to be promising towards mitigating water stress and energy consumption. However, those saline wastewaters contain nutrients (C, N, P) that need to be removed from the water before its discharge to the environment, to avoid severe environmental issues like hypoxia and eutrophication. Several biological nutrient removal processes have been developed to remove C, N and P, but the microorganisms responsible for the biological nutrient removal processes may be severely affected by salinity. Although the salinity effects on the C and N removal have been studied, the salinity effects on P removal are unclear. Only few studies have been conducted on the EBPR process but the outcomes from these studies are inconclusive and inconsistent. The objective of this part of the research was to get a better understanding of the effect of salinity on the microbial populations that prevail in the EBPR process and to develop a model describing the salinity effects on the metabolic processes of PAO and GAO. The specific research questions are defined as follows:

1) What are the short-term salinity effects on the anaerobic metabolism of PAO and GAO?
2) What are the short-term salinity effects on the aerobic metabolism of PAO and GAO?
3) How can the salinity effects on the metabolism of PAO and GAO be quantitatively described in model equations?

To find answers to the specific research questions, PAO II and GAO cultures were enriched and exposed in short-term batch experiments to different salt concentrations. From the outcomes of the first part of this research, PAO II seemed to be the most adaptive organisms towards changes in the environmental conditions and therefore PAO II was selected for the execution of the saline experiments. From a fundamental perspective, the outcomes of this study provide more insight into (i) the salinity effects on the different microbial populations that prevail in EBPR systems and (ii) more specifically on their metabolic processes. An inhibition model was developed which supports the expansion of design and operational guidelines for biological nutrient removal systems treating saline wastewaters.

1.5.3 Outline

Chapter 1 provides background information about the EBPR process and introduces the topic of this study and discusses the relevance of this research project. In Chapter 2, the functional differences in the anaerobic HAc-uptake kinetics and stoichiometry observed in short-term experiments of EBPR cultures, highly enriched with specific PAO clades (I, II and a mix of I and II). In Chapter 3 describes the effect of the storage polymers on the anaerobic stoichiometry and kinetics of a highly enriched PAO II culture assessed during long-term performance over a broad range of poly-P/active biomass values. The results were compared to the results of a previous study in literature that seemed to be conducted with a PAO I dominated culture. In Chapter 4 it was assessed if certain PAO clades could be enriched from activated sludge under phosphate limiting conditions to verify the findings of Chapter 2 and 3. In Chapter 5, the denitrification pathways of a highly enriched PAO I culture were assessed and the results were compared to literature to verify if functional differences exist in the denitrification pathways of PAO I and II.

Chapter 6 describes the short term salinity effects on the anaerobic metabolism (kinetics and stoichiometry) of highly enriched PAO and GAO cultures. In addition, an inhibition model was developed describing the salinity effects on the different anaerobic metabolic processes. In Chapter 7 the short term effects of salinity on the aerobic metabolism (kinetics and stoichiometry) of a highly enriched PAO culture were investigated. Also in this study an inhibition model was developed, describing the salinity effects on the different aerobic metabolic processes. Due to issues with the GAO enrichment, it was impossible to assess the salinity effects on the aerobic metabolism of a highly enriched GAO culture.

In chapter 8 the main findings and implications of this research are discussed as well as directions for future research.

1.6 References

Abu-Ghararah Z.H., Sherrard J.H. (1993) Biological nutrient removal in high salinity wastewaters. Journal of Environmental Science and Health 28(3): 559-613

Acevedo B., Oehmen A., Carvalho G., Seco A., Borras L., Barat R., (2012) Metabolic shift of polyphosphate-accumulating organisms with different levels of poly-phosphate storage. Water Research 46, 1889-1900

Barnard J.L. (1975) Biological nutrient removal without addition of chemicals. Water Research 9 (5-6), 458-490.

Barnard J.L. (1976) A review of biological phosphorus removal in the activated sludge process. Water SA 2, 136-144

Bohnert H.J. (1995) Adaptations to environmental stresses.Plant Cell 7: 1099-1111.

Bond, P.L., Hugenholtz, P., Keller, J., Blackall, L.L., 1995. Bacterial community structures of phosphate-removing and nonphosphate- removing activated sludges from sequencing batch reactors. Appl. Environ. Microbiol. 61 (5), 1910–1916.

Bond, P.L., Erhart, R., Wagner, M., Keller, J., Blackall, L.L., (1999). Identification of some of the major groups of bacteria in efficient and nonefficient biological phosphorus removal activated sludge systems. Appl. Environ. Microbiol. 65 (9), 4077–4084.

Brdjanovic, D., Van Loosdrecht, M. C. M., Hooijmans, C. M., Mino, T., Alaerts, G. J., Heijnen, J. J. (1998). Effect of polyphosphate limitation on the anaerobic metabolism of phosphorus-accumulating microorganisms. Applied microbiology and biotechnology, 50(2), 273-276.

Brown A.D. (1976) Microbial water stress. Bactcriol. Rev. 41: 803- 846.

Brown A.D. (1990) Microbial water stress physiology; Principles and perspectives

Burow, L.C., Kong, Y.H., Nielsen, J.L., Blackall, L.L., Nielsen, P.H., 2007. Abundance and ecophysiology of Defluviicoccus spp., glycogen accumulating organisms in full-scale wastewater treatment processes. Microbiology 153, 178–185.

Burow, L. C., Mabbett, A. N., McEwan, A. G., Bond, P. L., & Blackall, L. L. (2008). Bioenergetic models for acetate and phosphate transport in bacteria important in enhanced biological phosphorus removal. *Environmental microbiology, 10*(1), 87-98.

Castle A.M. Macnab R.M. Shulman R.G. (1986) Coupling between the Sodium and Proton Gradients in Respiring *Escherichia coli* Cells Measured by 23Na and 31P Nuclear Magnetic Resonance. The Journal of Biological Chemistry 261(17):7797-7806

Carvalho G., Lemos P.C., Oehmen A., Reis M.A.M. (2007) Denitrifying phosphorus removal: linking the process performance with the microbial community structure. Water Research 41(19), 4383, 4396

Caumette P., Cohen Y., Mathcron R. (1991) Isolation and characterization of Desulfovibrio halophilus sp. nov., a halophilic sulfate-reducing bacterium isolated from Solar Lake (Sinai). Syst. Appl. Microbiol. 14: 33- 38.

Comeau Y., Hall K.J., Hancock R.E.W., Oldham W.K. (1986) Biochemical-model for enhanced biological phosphorus removal. Water Res. 20 (12), 1511–1521.

Crocetti, G.R., Hugenholtz, P., Bond, P.L., Schuler, A., Keller, J., Jenkins, D., Blackall, L.L., 2000. Identification of polyphosphate- accumulating organisms and design of 16S rRNAdirected probes for their detection and quantitation. Appl.Environ. Microbiol. 66 (3), 1175–1182.

Crocetti, G.R., Banfield, J.F., Keller, J., Bond, P.L., Blackall, L.L., 2002. Glycogen-accumulating organisms in laboratory-scale and full-scale wastewater treatment processes. Microbiology 148, 3353–3364.

Cui Y., Peng C., Peng Y., Ye L. (2009) Effects of Salt on Microbial Populations and Treatment Performance in Purifying Saline Sewage Using the MUCT Process. Clean 37(8): 649-656

Dincer A. R., Kargi F. (1999) Salt Inhibition of Nitrification and Denitrification in Saline. *Environmental Technology* 20(11), 1147-1153

Fahim F.A., Fleita D.H. Ibrahim A.M., El-Dars F.M.S. (2000) Evaluation of some methods for fish canning wastewater treatment. Water, Air, ans Soil Pollution 127: 205-226

Filipe C.D.M., Daigger G.T., Grady C.P.L. (2001a) Effects of pH on the aerobic metabolism of phosphate-accumulating organisms and glycogen-accumulating organisms. Water Environ Res 73(2):213-222.

Filipe CDM, Daigger GT, Grady Jr CPL (2001b) A metabolic model for acetate uptake under anaerobic conditions by glycogen-accumulating organisms: stoichiometry, kinetics and effect of pH. Biotechnol Bioeng 76(1):17-31.

Flowers J.J., He S., Yilmaz S., Noguera D.R., McMahon K.D. (2009) Denitrification capabilities of two biological phosphorus removal sludges dominated by different ' Candidatus *Accumulibacter*' clades. Environmental Mivrobiology Reports 1(6): 583-588

Foekema H., van Thiel L., Lettinga B. (2008), *Watergebruik thuis 2007*. TNS NIPO

Galinski E.A., Truper H.G. (1994) Microbial behavior in salt-stressed ecosystems. FEMS Microbiology Reviews 15: 95-108

Galinski E.A. (1995) Osmoadaptation in bacteria. Adv. Microb.Physiol. 37, 273^328.

Gilmour D. (1990) halotolerant and halophilic microorganisms. In: Microbiology of Extreme Environments (Ed wards, C., Ed.), pp. 147-177. University Press, Milton Keynes.

Gonzaliez J.F., Civit E.M., Lupin H.M. (1983) Composition of fish filleting wastewater. Water SA 9(2): 49-56

Gu, A.Z., Saunders, A.M., Neethling, J.B., Stensel, H.D., Blackall, L., 2005. In: WEF (Ed.) Investigation of PAO and GAO and their effects on EBPR performance at full-scale wastewater treatment plants in US, October 29–November 2, WEFTEC, Washington, DC, USA.

Hascoet M.C. Florentz M., Granger P. (1985) Influence of nitrates on biological phosphorus removal from wastewater. Water Sa 11, 1-8

He, S., Gu, A.Z., McMahon, K.D., 2005. The role of Rhodocyclus-like organisms in biological phosphorus removal: factors influencing population structure and activity. In: Water Environment Federation Technical Exhibition and Conference (WEFTEC) 2005, October 29–November 2, Washington, DC, USA.

He S.,, Gall D. L., McMahon* K D. (2007) *"Candidatus* Accumulibacter" Population Structure in Enhanced Biological Phosphorus Removal Sludges as Revealed by Polyphosphate Kinase Genes_ applied and environmental microbiology 73(18), 5865-5874

Henze, M. (Ed.). (2008). Biological wastewater treatment: principles, modelling and design. IWA publishing.

Hesselmann, R.P.X., Werlen, C., Hahn, D., van der Meer, J.R., Zehnder, A.J.B., 1999. Enrichment, phylogenetic analysis and detection of a bacterium that performs enhanced biological phosphate removal in activated sludge. Syst. Appl. Microbiol. 22 (3), 454–465.

Hesselman, R.P.X., von Rummell, R., Resnick S.M., Hany, R., Zehnder A.J.B. (2000) Anaerobic Metabolism of Bacteria Performing Enhanced Biological Phosphate Removal. Water Research., 34, 3487

Hong C.C., Chan S.K., Shim H. (2007) Effect of chloride on biological nutrient removal from wastewater 2(3): 85-92

Intrasungkha N., Keller J., Blackall L.L. (1999) Biological nutrient removal efficiency in treatment of saline wastewater. Water Science and Technology 39(6): 183-190

Janssen, P. M. J., Meinema, K., & van der Roest, H. F. (Eds.). (2002) Biological phosphorus removal: manual for design and operation. IWA publishing.

Jiang X. (2004), Household Water Consumption Pattern in Beijing (Chinese version). Economical Information Centre - Department of Economic Research and Consultancy, Beijing.

Kargi, F., A. Uygur, (2005), Improved Nutrient Removal from Saline Wastewater in an SBR by *Halobacter* Supplemented Activated Sludge. *Envir. Eng. Sci.*, v. 22 (2), p. 170-176.

Kempf B., Bremer E. (1995) OpuA, an osmotically regulated binding protein-dependent transport system for the osmoprotectant glycine betaine in Bacillus subtilis. J Biol Chem 270: 16701-16713.

Kong, Y.H., Beer, M., Seviour, R.J., Lindrea, K.C., Rees, G.N., 2001. Structure and functional analysis of the microbial community in an aerobic: anaerobic sequencing batch reactor (SBR) with no phosphorus removal. Syst. Appl. Microbiol. 24 (4), 597–609.

Kong, Y.H., Ong, S.L., Ng, W.J., Liu, W.T., (2002b). Diversity and distribution of a deeply branched novel proteobacterial groupfound in anaerobic–aerobic activated sludge processes. Environ.Microbiol. 4 (11), 753–757.

Kong, Y.H., Nielsen, J.L., Nielsen, P.H., 2004. Microautoradiographic study of Rhodocyclus-Related polyphosphate-accumulating bacteria in full-scale enhanced biological phosphorus removal plants. Appl. Environ. Microbiol. 70 (9), 5383–5390.

Kong, Y.H., Xia, Y., Nielsen, J.L., Nielsen, P.H., 2006. Ecophysiology of a group of uncultured Gammaproteobacterial glycogenaccumulating organisms in full-scale enhanced biological phosphorus removal wastewater treatment plants. Environ. Microbiol. 8 (3), 479–489.

Lanyi J.K. (1974) Salt-dependent properties of proteins from extremely halophilic bacteria. Bacteriol. Rev. 38: 272-290

Larsen H. (1986) Halophilic and halotolerant microorganisms - an overview and historical perspective. FEMS Microbiol. Rev. 39, 3-7.

Lefebvre O., Moletta R., (2006) Treatment of organic pollution in industrial saline wastewater: A literature review. Water Research 40: 3671-3682.

Lemos, P. C., Dai, Y., Yuan, Z., Keller, J., Santos, H., & Reis, M. A. (2007). Elucidation of metabolic pathways in glycogen-accumulating organisms with in vivo13C nuclear magnetic resonance. Environmental microbiology, 9(11), 2694-2706.

Leung R.W.K., Li D.C.H., Yu W.K., Chui H.K., Lee T.O., Van Loosdrecht M.C.M., Chen G.H. (2012) Integration of seawater and grey water reuse to maximize alternative water resource for coastal areas: The case of the Hong Kong International Airport. Water Science and Technology 65(3): 410-417

Levin, G. V., & Shapiro, J. (1965). Metabolic uptake of phosphorus by wastewater organisms. Journal (Water Pollution Control Federation), 800-821.

Liu W.T., Mino T., Nakamura K., Matsuo T. (1994) Role of glycogen in acetate uptake and polyhydroxyalkanoate synthesis in anaerobic-aerobic activated sludge with a minimized polyphosphate content. J Ferment Bioeng 77(5):535-540.

Lopez-Vazquez C.M., Hooijmans C.M., Brdjanovic D., Gijzen H.J., van Loosdrecht M.C.M. (2007) A Practical Method for the Quantification of Phosphorus- and Glycogen-Accumulating Organisms Populations in Activated Sludge Systems. *Water Environment Research* 79(13): 2487-2498

Lopez-Vazquez, C. M., Hooijmans, C. M., Brdjanovic, D., Gijzen, H. J., & van Loosdrecht, M. C. (2008). Factors affecting the microbial populations at full-scale enhanced biological phosphorus removal (EBPR) wastewater treatment plants in The Netherlands. Water research, 42(10), 2349-2360.

Lopez-Vazquez C.M., Oehmen A., Hooijmans C.M., Brdjanovic D., Gijzen H.J., Yuan Z., van Loosdrecht M.C.M. (2009) Modelling the PAO-GAO competition: Effects of carbon source, pH and temperature. Water Research 43(2) 450-462.

Mariangel L., Aspe E., Marti M.C., Roeckel M. (2008) The effect of sodium chloride on the denitrification of saline fishery wastewaters. *Environmental Technology*, 29(8), 871-879

Martín H.G., Ivanova N., Kunin V., Warnecke F., Barry K.W., McHardy A.C., Yeates C., He S., Salamov A.A., Szeto E., Dalin E., Putnam N.H., Shapiro H.J., Pangilinan J.L., Rigoutsos I., Kyrpides N.C., Blackall L.L., McMahon K.D., Hugenholtz P. (2006) Metagenomic analysis of two enhanced biological phosphorus removal (EBPR) sludge communities. *Nature Biotechnology* **24**, 1263 - 1269

Maurer M., Gujer W., Hany R., Bachmann S. (1997) Intracellular Carbon Flow in Phosphorus Accumulating Organisms from Activated Sludge Systems. Water research 31(4), 907-917

Mayer P.W., DeOreo W.B., Opit, E.M., Kiefer J.C., Davis W.Y., Dziegielewski B., Nelson J.O. (1999), Residential end uses of water study. AWWA Research Foundation and American Water Works Association.

Measures J.C. (1975) Role of amino acids in osmoregulation of non-halophilic bacteria. Nature 257: 398-400

Metcalf and Eddy, Inc. (2003) Wastewater Engineering - Treatment and Reuse, 4th Edn. New York, USA: Mc Graw Hill.

Meyer, R.L., Saunders, A.M., Blackall, L.L., 2006. Putative glycogenaccumulating organisms belonging to Alphaproteobacteria identified through rRNA-based stable isotope probing. Microbiology152, 419–429.

Mino T., Arun V., Tsuzuki Y., Matsuo T., (1987) Effect of phosphorus accumulation on acetate metabolism in the biological phosphorus removal process. In: Ramadori, R. (Ed.), Biological Phosphate Removal from Wastewaters, Advances in Water Pollution Control, vol 4. Pergamon Press, Oxford (1987), pp. 27-38

Mino T., van Loosdrecht M.C.M., Heijnen J.J. (1998) Microbiology and Biochemistry of the Enhanced Biological Phosphate Removal Process. Water Research 32(11): 3193-3207

Moussa M.S., Sumanasekera D.U., Ibrahim S.H., Lubberding H.J., Hooijmans C.M., Gijzen H.J., Van Loosdrecht, M.C.M. (2006) Long Term effects of Salt on Activity, Population Structure and Floc Characteristics in Enriched Bacterial Cultures of Nitrifiers. *Water Research*, 40, 1377-1388

Nielsen, A.T., Liu, W.T., Filipe, C., Grady, L., Molin, S., Stahl, D.A., 1999. Identification of a novel group of bacteria in sludge from a deteriorated biological phosphorus removal reactor. Appl. Environ. Microbiol. 65 (3), 1251–1258.

Oehmen, A., Yuan, Z., Blackall, L.L., Keller, J., 2004. Short-term effects of carbon source on the competition of polyphosphate accumulating organisms and glycogen accumulating organisms. Water Sci. Technol. 50 (10), 139–144.

Oehmen A., Lemos P.C., Carvalho G., Yuan Z., Keller J., Blackall L.L., Reis M.A.M. (2007) Advances in enhanced biological phosphorus removal: From micro to macro scale. Water Research 41: 2271-2300

Ollivier B., Hatchikian C.E., Prensier G., Guezennec J., Garcia J.L. (1991) Desulfohalobium retbaense gen. nov., sp. nov., a halophilic sulfate-reducing bacterium from sediments of a hypersaline lake in Senegal. Int. J. Syst. Appl. Bacteriol. 41: 74- 81.

Oren A. (1991) Estimation of the contribution of archaebacteria and eubacteria to the bacterial biomass and activity in hypersaline ecosystems: novel approaches. In: General and Applied Aspects of Halophilic Bacteria, Rodgriguez-Valera, Ed.), pp. 25-31. Plenum Press, New York and London.

Oren A. (1999) Bioenergetic aspects of halophilism. Microbiology and Molecular Biology Reviews, June 1999, 334-348

Orhon D., Tasli R., Sozen S. (1999) Experimental basis of activated sludge treatment for industrial wastewaters - the state of the art. Water Science and Technology 40(1): 1-11

Panswad T., Anan C. (1999) Impact of high chloride wastewater on an anaerobic/anoxic/aerobic process with and without inoculation of chloride acclimated seeds. Water Research 33(5): 1165-1172

Pereira, H., lemos P.C. Reis M.A.M., Crespo J.P.S.G. Carrondo M.J.T. Santos H. (1996) Model for Carbon Metabolism in Biological Phosphorus Removal Processes Based on In Vivo C13-NMR Labelling Experiments. Water Research. 30, 2128

Peterson SB., Warnecke F., Madejska J., McMahon K.D., and Hugenholtz P., (2008) Environmental distribution and population biology of *Candidatus* Accumulibacter, a primary agent of biological phosphorus removal Environmental Microbiology 10(10), 2692-2703

Pijuan M., Ye L., Yuan Z. (2010) Free nitrous acid inhibition on the anaerobic metabolism of poly-phosphate accumulating organisms. *Water research* 44(20), 6063-6072

Pronk M., Bassin J.P., de Kreuk M.K., Kleerebezem R., van Loosdrecht M.C.M. (2013) Evaluating the main side effects of high salinity on aerobic granular sludge. Appl Mirobiol Biotechnol 98: 1339-1348

Record Jr. M.T., Courtenay E.S., Cayley D.S., Guttman H.J. (1998) Responses of Escherichia coli to osmotic stress: large changes in amounts of cytoplasmic solutes and water. Trends Biochem. Sci.23: 143-148.

Roberts M.F. (2000) Osmoadaptation and osmoregulation in archaea. Front Biosci 5: 796-812.

Roesler M., Muller V. (2001) Osmoadaptation in bacteria and archae: common principles and differences. Environmental Microbiology 3(12): 743-754

Russel N.J. (1989) Adaptive modifications in membranes of halotolerant and halophilic microorganisms. J. Bioenerg. Biomembr. 21: 93- 113.

Satoh H., Mino T., Matsuo T., (1992) Uptake of organic substrates and accumulation of polyhydroxyalkanoates linked with glycolysis of intracellular carbohydrates under anaerobic conditions in the biological excess phosphate removal processes. Water Sci. Technol. 26(5–6): 933–942.

Saunders, A. M., Mabbett, A. N., McEwan, A. G., & Blackall, L. L. (2007). Proton motive force generation from stored polymers for the uptake of acetate under anaerobic conditions. FEMS microbiology letters, 274(2), 245-251.

Schuler A.J., Jenkins D. (2003) Enhanced Biological Phosphorus Removal from Wastewater by Biomass with Different Phosphorus Contents, Part 1: Experimental Results and Comparison with Metabolic Models. Water Environment Research 75(6), 485-498

Sleator R.D., Hill C. (2001) Bacterial Osmoadaptation: the role of osmolytes in bacterial stress and virulence. FEMS Microbiology reviews 26: 49-71

Srinath, E. G., Sastry, C. A., & Pillai, S. C. (1959). Rapid removal of phosphorus from sewage by activated sludge. Experientia, 15(9), 339-340.Sutton G.C., Russel N.J, Quinn P.J. (1991) The effect of salinity on the phase behaviour of total lipid extracts and binary mixtures of the major phospholipids isolated from a moderately halophilic eubaclerium. Biochim. Biophys. Acta 1061, 235-246.

Tang S.L., Yue D.P.T., Li X.Z. (2006) Comparison of engineering costs of raw freshwater, reclaimed water and seawater for toilet flushing in Hong Kong. Water and Environment Journal 20(4): 240-247

Tindall B.J. (1988) Prokaryotic life in the alkaline, saline, athalassic environment. In: Halophilic Bacteria (Rodri- guez-Valera, F., Ed.), pp. 31-67. CRC Press, Boca Raton, FL.

Truper H.G., Galinski E.A. (1986) Concentrated brines as habitats for microorganisms. Experientia 42: 1182-1187.

Satoh H, Mino T, Matsuo T (1994) Deterioration of enhanced biological phosphorus removal by the domination of microorganisms without polyphosphate accumulation. Water Science and Technology, 30(6):203-211.

Saunders, A.M., Oehmen, A., Blackall, L.L., Yuan, Z., Keller, J., 2003. The effect of GAO (glycogen accumulating organisms) on anaerobic carbon requirements in full-scale Australian EBPR (enhanced biological phosphorus removal) plants. Water Sci. Technol. 47 (11), 37–43.

Slater F.R., Johnson C.R., Blackall L.L., Beiko R.G., Bond P.L. (2010) Monitoring associations between clade-level variation, overall community structure and ecosystem function in enhanced biological phosphorus removal(EBPR) systems using terminal-restriction fragment length polymorphism (T-RFLP). Water Research 2010: 1-16

Smolders G.J.F., van der Meij J., van Loosdrecht M.C.M. and Heijnen J.J. (1994) Model of the Anaerobic Metabolism of the Biological Phosphorus Removal Process: Stoichiometry and pH influence. Biotechnology and Bioengineering (43), 461-470

Uygur A., Kargi F. (2004) Salt Inhibition on biological nutrient removal from saline wastewater in a sequencing batch reactor. Enzyme and Mic. Tech., 34: 313-318

Uygur A. (2006) Specific nutrient removal rates in saline wastewater treatment using sequencing batch reactor. Proc. Bio. 41(1): 61-66

Van den Brand, T. P. H. (2014). Sulphate reducing bacteria in wastewater treatment. TU Delft, Delft University of Technology.

Vredenbregt L.H.J., Potma A.A., Nielsen K., Kristensen G.H., Sund C. (1997) Fluid bed biological nitrification and denitrification in high salinity wastewater. Water Science and Technology 36 (1), 93–100.

Wentzel M.C., Dold P.L.., Ekama G.A. and Marais G.v.R. (1985) Kinetics of biological phosphorus release. Water Science and Technology 17, 57-71

Wentzel M.C., Lotter L.H., Loewenthal R.E., Marais G.v.R. (1986) Metabolic behaviour of Acinetobacter spp. in enhanced biological phosphorus removal—a biochemical model. Water Sa 12, 209–224.

Wentzel, M. C., Dold, P. L., Loewenthal, R. E., Ekama, G. A., & Marais, G. V. R. (1987). Experiments towards establishing the kinetics of biological excess phosphorus removal. In Biological Phosphate Removal from Wastewaters: Proceedings of an Iawprc Specialized Conference Held in Rome, Italy, 28-30 September, 1987 (Vol. 4, p. 79). Pergamon.

Wentzel, M. C., Loewenthal, R. E., Ekama, G. A., & Marais, G. V. R. (1988). Enhanced polyphosphate organism cultures in activated sludge systems- Part 1: Enhanced culture development. Water S. A., 14(2), 81-92.

Wong, M.T., Tan, F.M., Ng, W.J., Liu, W.T., 2004. Identification and occurrence of tetrad-forming Alphaproteobacteria in anaerobic– aerobic activated sludge processes. Microbiology 150, 3741–3748.

Wong, M.T., Mino, T., Seviour, R.J., Onuki, M., Liu, W.T., 2005. In situ identification and characterization of the microbial community structure of full-scale enhanced biological phosphorous removal plants in Japan. Water Res. 39 (13), 2901–2914.

Wu G., Guan Y., Zhan X. (2008) Effect of salinity on the activity, settling and microbial community of activated sludge in a sequencing batch reactors treating synthetic saline wastewater. Water Science and Technology 58(2): 351-358

Yancey P.H., Clark M.E., Hand S.C., Bowlus R.D., Somero G.N. (1982) Living with water stress : evolution of osmolyte systems. Science 217: 1214-1222.

Yancey P.H. (1994) Compatible and counteracting solutes. In: Cellular and Molecular Physiology of Cell Volume Regulation (Strange, K., Ed.), pp. 81-82. CRC Press, Boca Raton, FL

Ye L., Pijuan M., Yuan Z. (2010) The effect of nitrous acid on the anabolic and catabolic processes of glycogen accumulating organicms. Water research 44: 2901-2909

Zeng, R., Yuan, Z., van Loosdrecht, M., and Keller, J. (2002). Proposed modifications to metabolic model for glycogen-accumulating organisms under anaerobic conditions. Biotechnology and bioengineering, 80(3), 277-279.

Zeng R.J., van Loosdrecht M.C.M., Yuan Z.G., Keller J. (2003a). Metabolic model for glycogen-accumulating organisms in anaerobic/aerobic activated sludge systems. Biotechnol.Bioeng. 81 (1), 92–105.

Zeng, R.J., Yuan, Z., Keller, J., 2003b. Model-based analysis of anaerobic acetate uptake by a mixed culture of polyphosphate-accumulating and glycogen-accumulating organisms. Biotechnol. Bioeng. 83 (3), 293–302.

Zhilina T.N., Zavarzin G.A. (1990) Extremely halophilic, methylotrophic, anaerobic bacteria. FEMS Microbiol. Rev. 87, 315 322.

Zhou Y., Pijuan M., Zeng R.J., Lu H., Yuan Z., (2008) Could polyphosphate-accumulating organisms (PAO) be glyccogen-accumulating organisms (GAO)? Water Research 42, 2361-2368

Zilles, J.L., Peccia, J., Kim, M.W., Hung, C.H., Noguera, D.R., 2002. Involvement of Rhodocyclus-related organisms in phosphorus removal in full-scale wastewater treatment plants. Appl. Environ. Microbiol. 68 (6), 2763–2769.

2

Accumulibacter clades Type I and II performing kinetically different glycogen-accumulating organisms metabolisms for anaerobic substrate uptake

Abstract

The anaerobic acetate (HAc) uptake stoichiometry of polyphosphate-accumulating organisms (PAO) in enhanced biological phosphorus removal (EBPR) systems has been an extensive subject of study due to the highly variable reported stoichiometric values (e.g. anaerobic P-release/HAc-uptake ratios ranging from 0.01 up to 0.93 P-mol/C-mol). Often, such differences have been explained by the different applied operating conditions (e.g. pH) or occurrence of glycogen-accumulating organisms (GAO). The present study investigated the ability of biomass highly enriched with specific PAO clades ('*Candidatus* Accumulibacter phosphatis' Clade I and II, hereafter PAO I and PAO II) to adopt a GAO metabolism. Based on long-term experiments, when Poly-P is not stoichiometrically limiting for the anaerobic VFA uptake, PAO I performed the typical PAO metabolism (with a P/HAc ratio of 0.64 P-mol/C-mol); whereas PAO II performed a mixed PAO-GAO metabolism (showing a P/HAc ratio of 0.22 P-mol/C-mol). In short-term batch tests, both PAO I and II gradually shifted their metabolism to a GAO metabolism when the Poly-P content decreased, but the HAc-uptake rate of PAO I was 4 times lower than that of PAO II, indicating that PAO II has a strong competitive advantage over PAO I when Poly-P is stoichiometrically limiting the VFA uptake. Thus, metabolic flexibility of PAO clades as well as their intrinsic differences are additional factors leading to the controversial anaerobic stoichiometry and kinetic rates observed in previous studies. From a practical perspective, the dominant type of PAO prevailing in full-scale EBPR systems may affect the P-release processes for biological or combined biological and chemical P-removal and recovery and consequently the process performance.

Adapted from:

Welles, L., Tian, W. D., Saad, S., Abbas, B., Lopez-Vazquez, C. M., Hooijmans, C. M., van Loosdrecht M.C.M., Brdjanovic, D. (2015). Accumulibacter clades Type I and II performing kinetically different glycogen-accumulating organisms metabolisms for anaerobic substrate uptake. *Water research, 83,* 354-366.

2.1 Introduction

The enhanced biological phosphorus removal (EBPR) process is often implemented in wastewater treatment activated sludge processes through the enrichment of polyphosphate-accumulating organisms (PAO). PAO are able to take up and store phosphate as intracellular polyphosphate (Poly-P), leading to P removal from the bulk liquid by wastage of activated sludge. To favor the development of PAO, sludge is cycled through anaerobic and aerobic stages (Barnard *et al.*, 1975). In the anaerobic phase, PAO take up volatile fatty acids (VFA), such as acetate (HAc) and propionate (HPr), and store them intracellularly as poly-β-hydroxyalkanoates (PHA) (Wentzel *et al.*, 1985, Comeau *et al.*, 1986, Mino *et al.*, 1987). Energy is mainly generated by the cleavage of intracellular Poly-P and release of phosphate over the cell membrane. For the production of reducing power and additional required energy, glycolysis of intracellularly stored glycogen has been proposed as the main pathway (Mino *et al.*, 1987; Smolders *et al.*, 1994). In the aerobic (or anoxic) phase, PAO utilize the anaerobically stored PHA as carbon and energy source for growth, maintenance, and recovery of glycogen and Poly-P pools. Once phosphate is stored within the PAO cells, possibilities exist to recover the phosphate in side-stream processes by stimulating the P-release through VFA addition.

In spite of several studies executed with enriched EBPR cultures, the stoichiometry and kinetics of EBPR anaerobic conversions are still controversial. For instance, reported anaerobic P-release/HAc-uptake ratios vary from 0.01 up to 0.93 P-mol/C-mol (Pereira *et al.*, 1996; Hesselman *et al.*, 2000; Smolders *et al.*, 1994; Wentzel *et al.*, 1987; Kisoglu *et al.*, 2000; Schuler and Jenkins 2003). These differences partially occur due to the influence of pH on the P/HAc ratio (Smolders *et al.* 1994). Barat *et al.* (2006, 2008) associated a decrease in the anaerobic P/HAc ratio with an increase in calcium concentration in the influent. From a microbiological perspective, the co-existence with glycogen accumulating organisms (GAO) is another potential factor affecting the observed stoichiometry (Mino *et al.*, 1998, Schuler and Jenkins 2003). GAO solely rely on glycogen as energy source for VFA uptake and, consequently, the glycogen consumption and PHA production per VFA uptake (mainly as PHV) are higher when GAO are present, resulting in lower P/HAc ratios (Schuler and Jenkins. 2003). Nonetheless, EBPR cultures have also exhibited in short-term experiments a (partial) GAO metabolism under conditions where Poly-P was stoichiometrically limiting for the VFA uptake, i.e. when there was not enough poly-P available for the organisms to take up VFA, using a typical PAO metabolism (Brdjanovic *et al.*, 1998, Hesselman et al., 2000; Zhou *et al.*, 2008; Erdal *et al.*, 2008; Acevedo *et al.*, 2012). In two of these studies, microbial community analysis demonstrated that these cultures were highly enriched with PAO, confirming that at least some PAO clades were able to perform a metabolic shift (Zhou et al., 2008; Acevedo et al., 2012).

In long-term studies, Liu *et al.* (1997), Kong *et al.* (2002)and Schuler and Jenkins (2003) demonstrated that a (partial) GAO metabolism tends to be present under P-limiting conditions, i.e. phosphate is fully removed by the end of the aerobic phase. A gradual shift from a PAO to a GAO metabolism was observed when the influent P/C ratios and therefore the intracellular Poly-P content of the biomass decreased (Liu et at., 1997; Schuler and Jenkins, 2003; Kong et al., 2002) Still, those results are inconclusive. Due to lack of data on the microbial communities, it remains unclear whether there was a metabolic shift (Schuler and Jenkins, 2003) and/or a shift in the microbial community structure from a PAO- to a GAO-dominated culture (Liu et at., 1997; Schuler and Jenkins, 2003). Furthermore, at similar Poly-P contents the stoichiometric values are not consistent (e.g. P/HAc ratios vary from 0.2 to 0.5 P-mol/C-mol at Total-P/TSS ratios of 0.08 mg/mg and between 0.5 and 0.9 P-mol/C-mol at a total-P/TSS ratio of 0.14) (Cech and Hartman, 1993; Liu et al., 1997; Kisoglu et al., 2000; Sudiana et al., 1999; Schuler and Jenkins, 2003). This suggests that, besides the Poly-P content, the existence of different microbial communities may have affected the stoichiometry. In this regard, Kong et al. (2002) observed a shift from a β-*Proteobacteria* dominated culture (the subdivision to which most PAO belong) at a high influent P/C ratio to a community dominated by *a-Proteobacteria* and *γ-Proteobacteria* (the subdivisions to which most

GAO belong) at a low influent P/C ratio, which could explain the observed changes in the anaerobic stoichiometry. A shift in the abundance of different PAO clades as a possible cause for changes in the anaerobic stoichiometry was not discarded (Kong et al., 2002). Using both the 16SrRNA gene and the poly-phosphate kinase gene (ppk1) as a genetic marker, it was revealed that *Accumulibacter* was organized into two main clades, *Candidatus* Accumulibacter phosphatis clade I (PAO I) and *Candidatus* Accumulibacter phosphatis clade II (PAO II) (He *et al.*, 2007; Peterson *et al.*, 2008). Interestingly, Carvalho et al. (2007), Flowers et al. (2009) and Slater et al. (2010) observed intrinsic differences in the physiological and morphological properties of PAO clades and Acevedo et al. (2012) noticed a shift from a culture dominated by PAO Clade Type I at a high biomass P-content to a mixed culture of PAO Clades Type I and II (hereafter PAO I and PAO II) at a low biomass P-content. These observations support the hypothesis that the different PAO clades can use different metabolic pathways for the anaerobic conversions being the phosphate availability a factor affecting the competition between PAO clades and thereby influencing the anaerobic stoichiometry.

Thus, the objectives of this study were (i) to assess if certain PAO clades exhibit a (partial) GAO metabolism during long-term operation under conditions were the influent phosphate concentrations are still limiting (full P-removal is obtained in the system, but where the intracellular Poly-P levels were not stoichiometrically limiting for VFA uptake; (ii) to evaluate if both PAO clades Type I and II are able to shift to a GAO metabolism in short-term tests at different Poly-P levels and (iii) to study the stoichiometry and VFA uptake rates of PAO clades Type I and II during Poly-P depletion. This can contribute to elucidate the cause of the variations in in the PAO and GAO metabolisms observed in past studies, help to improve the existing metabolic models and, ultimately, lead to a better design and operation of EBPR and phosphate recovery processes.

2.2 Material and Methods

2.2.1 Enrichment of the PAO cultures

2.2.1.1 Operation of sequencing batch reactors (SBR)
Three PAO cultures enriched in three similar 2.5 L double-jacketed laboratory sequencing batch reactors (SBR), namely SBR-L, SBR-S and SBR-W, were used in this study. Activated sludge from a municipal wastewater treatment plant (Hoek van Holland, The Netherlands) was used as inoculum. The SBR were operated, following similar procedures as those described by (Smolders *et al.*, 1994). With the aim to enrich different PAO clades in the different systems, some modifications were made in the operational conditions of each system on a trial base, as selective conditions for PAO clades have not been reported in previous studies. The differences in operational factors between the SBR are listed in Table 2.1. Further details regarding the operational conditions of SBR-L, SBR-S and SBR-W can be found elsewhere (Welles et al., 2014, Saad et al., in preparation).

Reactors	SRT (d)	T (°C)	pH	VFA	Influent P (mgP/L)	Influent COD (mgCOD/L)	P/C ratio (P-mol/C-mol)	Total cycle (h)	Anaerobic phase (h)	Aerobic phase (h)	Settling time (h)
SBR-L	8	20±1	7.0±0.05	HAc	15	400	0.038	6.00	2.25	2.25	1.50
SBR-W	32	10±1	7.0±0.10	HAc	15	300	0.051	6.00	2.00	3.00	1.00
SBR-S	8	20±1	7.6±0.05	HAc (75%), HPr (25%)	25	400	0.066	6.00	2.25	2.25	1.50

Table 2.1 Operational parameters of SBR-L, SBR-W and SBR-S.

2.2.1.2 Synthetic medium

All SBR were fed with synthetic media. During the execution of this study, the influent phosphate concentration fed to SBR-L was 15 mg PO_4-P/L (0.48 P-mmol/L, 67 mg$NaH_2PO_4.H_2O$), while the influents of SBR-W and SBR-S contained 15 mgPO_4-P/L (0.48 P-mmol/L, 67 mg$NaH_2PO_4.H_2O$) and 25 mgPO_4-P/L (0.80 P-mmol/L, 112 mg$NaH_2PO_4.H_2O$), respectively. Besides the influent phosphorus concentration, different compositions and concentrations of volatile fatty acids (VFA)(such as acetate - HAc- and propionate -HPr-) were fed to each reactor. The influent of SBR-L contained 373 mg HAc/L (12.6 C-mmol/L, 400 mg COD/L, 860 mg NaAc·$3H_2O$), leading to an influent P/C ratio of 0.038 (P-mol/C-mol); while the HAc concentration fed to SBR-W was limited to 280 mg HAc/L (9.5 C-mmol/L, 300 mg COD/L, 645 mg NaAc·$3H_2O$), resulting in an influent P/C ratio of 0.051 (P-mol/C-mol). The influent of SBR-S contained a 75%-to-25% mixture of acetate-to- propionate: 280 mg HAc/L (9.5 C-mmol/L, 300 mg COD/L, 645 mg NaAc·$3H_2O$) and 67 mg/L HPr (2.72 C-mmol/L, 100 mg COD/L, 6.675x10^{-2} ml $C_3H_6O_2$), leading to an influent P/C ratio of 0.066 (P-mol/C-mol). Further details regarding other macronutrients and trace elements can be found elsewhere (Smolders et al., 1994; Welles et al., 2014).

2.2.1.3 Monitoring of SBR

The performance of all SBR was regularly monitored by measuring orthophosphate (PO_4-P), acetate (HAc-C), mixed liquor suspended solids (MLSS) and mixed liquor volatile suspended solids (MLVSS). The (pseudo) steady-state condition in the reactors was confirmed through the daily determination of the aforementioned parameters as well as pH and DO.

2.2.1.4 Characterization of biomass activity

When both SBR reached steady-state conditions, cycle measurements were carried out to determine the biomass kinetic rates and stoichiometry of the anaerobic conversions. In addition to the above described parameters, poly-beta-hydroxyalkanoate (PHA) and glycogen were measured in the cycles. When relatively lower anaerobic P-release values were observed (like in SBR-L and SBR-W), K, Mg and Ca were also analyzed to assess whether any potential chemical precipitation occurred (which could partially mask biological P-release by PAO). K was measured by atomic emission and Mg and Ca by atomic absorption spectrometry using a Perkin Elmer Analyst 200 Atomic Absorption Spectrometer.

2.2.1.5 Identification of microbial populations

An estimation of the degree of enrichment of the bacterial populations of interest (PAO I, PAO II and GAO) was undertaken via fluorescence in situ hybridization (FISH) analyses, following the procedure described by Winkler et al. (2011). All bacteria were targeted by the EUB338mix (general bacteria probe) (Amann et al., 1990; Daims et al., 1999). β-proteobacteria and γ-proteobacteria were identified with BET42 and GAM42a probes, respectively (Manz et al., 1992). 'Candidatus Accumulibacter phosphatis' was targeted by PAOMIX probe (mixture of probes PAO462, PAO651 and PAO846) (Crocetti et al., 2000) whereas GAOMIX probe (mixture of probes GAOQ431 and GAOQ989) (Crocetti et al., 2002) was used to target 'Candidatus Competibacter phosphatis'. PAO I (clade IA and other type I clades) and PAO II (clade IIA, IIC and IID) were targeted by the probes Acc-1-444 and Acc-2-444 (Flowers et al., 2009), respectively.

To identify the dominant microbial populations, 16S-rRNA gene denaturing gradient gel electrophoresis (DGGE) was applied. Samples were collected from all SBR systems at the end of each cycle. DNA extraction, PCR amplification, DGGE, band isolation, sequencing and identification of microorganisms were carried out according to the procedures described by Bassin et al. (2011). A phylogenetic tree was generated using maximum likelihood algorithm implemented into ARB.

2.2.2 Poly-P depletion tests

2.2.2.1 Batch activity tests

To gradually deplete the intracellular Poly-P content and assess the HAc uptake ability at low Poly-P levels, batch tests were performed in a similar way as described by Brdjanovic *et al.* (1998). After reaching pseudo steady-state conditions, batch experiments were performed in 0.5 L double-jacketed laboratory reactors at $20\pm0.5^{\circ}$C. 400 mL of mixed liquor were withdrawn at the end of the aerobic phase and transferred to the batch reactor. Biomass was allowed to settle and 200 mL of supernatant was removed prior to the start of the anaerobic feeding phase.

The batch tests consisted of 4 consecutive cycles for SBR-L and SBR-W, and 6 consecutive cycles for SBR-S (because of the higher P-content of the SBR-S biomass). The first cycles were designed to deplete the Poly-P content. The last cycles (4th cycle for SBR-L and SBR-W, and 6th cycle for SBR-S) were executed to investigate the PAO HAc uptake ability when the Poly-P levels were low or depleted. The first cycles were similar and consisted of: (i) an anaerobic feeding phase of 180 min; (ii) an anaerobic washing phase of 60 min to remove the phosphate released during the anaerobic feeding phase, the procedure of which is described in more detail in the next paragraph; (iii) an aerobic phase for recovery of the intracellular glycogen content of the biomass with a variable length; and, (iv) a 20 min settling phase followed by effluent removal (leaving 50% of the working volume). The only difference in the first cycles was the length of the aerobic phase. To speed up the Poly-P depletion process in the first cycles, the influent HAc concentrations were doubled (up to 25.2 C-mmol/L) compared to that of the parent SBR and the pH set at 7.5. Although these conditions enhance the anaerobic P-release when compared to the conditions applied in the SBR systems, they are commonly applied in during long-term laboratory studies with EBPR systems and have not been found to affect the integrity of the PAO cells. For comparison purposes, the last cycles (4th for SBR-L and SBR-W and 6th for SBR-S) were performed at pH 7.0 with a HAc concentration of 12.6 C-mmol/L and only comprised an anaerobic feeding phase.

In all batch tests, N_2 gas was sparged at a flow rate of 6 L/h to create the anaerobic conditions and compressed air at 12 L/h to create the aerobic phases. Biomass was constantly stirred at 500 rpm, except during the washing and settling phases when the stirrers were turned off. In the anaerobic washing phases, biomass was settled for 20 min followed by 50% volume withdrawal which was replaced by a mineral solution while N_2 was sparged. To increase the P-removal efficiency, after 5 min, the washing phase was repeated (20 min settling and 50% volume replacement with mineral solution). Thereafter, the corresponding aerobic phase continued with the air supply. The mineral solution used for the washing and aerobic phases was similar to the medium supplied to SBR-L, except for the absence of ortho-phosphate and HAc. Solutions were sparged with nitrogen gas prior to addition to the batch reactors to keep the anaerobic conditions. Since the settling ability of SBR-S biomass drastically decreased in the 5th cycle, the biomass was concentrated by centrifugation at 3500 rpm for 5 min.

2.2.2.2 Sampling

During all cycles, samples were collected for the determination of MLVSS, MLSS, PHA, glycogen, HAc and PO_4-P at the beginning and end of the anaerobic stages. In the last cycles, additional samples for HAc and PO_4-P determination were taken throughout the phase to determine their kinetic rates.

2.2.3 Analyses

MLSS and MLVSS and PO_4-P determinations were performed in accordance with Standard Methods (A.P.H.A., 1995). HAc was determined using a Varian 430-GC Gas Chromatograph (GC), equipped with a split injector (split ratio 1:10), a WCOT Fused Silica column with a FFAP-CB coating (25 m x 0.53mm x

$1\mu m$), and coupled to a FID detector. Helium gas was used as carrier gas. Temperature of the injector, column and detector was 200°C, 105°C and 300°C, respectively. Glycogen was determined according to the method described by Smolders *et al.* (1994) but with a digestion phase extended to 5 h. PHB and PHV content of freeze dried biomass was determined according to the method described by Smolders *et al.* (1994).

2.2.4 Active biomass

The active biomass concentration was determined as MLVSS excluding PHB, PHV and glycogen (active biomass = MLVSS – PHB – PHV – glycogen). Unbiodegradable particulate endogenous residue, shown to be another non-active biomass component of the MLVSS (Wentzel et al., 1988, 1989ab) was neglected for the sake of simplicity. The active biomass concentration was expressed in C-mol units by taking into account the experimentally determined composition of PAO ($CH_{2.09}O_{0.54}N_{0.20}P_{0.015}$) (Smolders et al., 1994).

2.2.5 Determination of kinetic and stoichiometric parameters

The kinetic rates of interest were the anaerobic HAc-uptake rate and the aerobic PO_4-uptake rate. The rates were expressed as maximum active biomass specific rates based on the HAc and PO_4-P profiles observed in the tests as described by Smolders *et al.* (1994b) and Brdjanovic et al., 1997). The stoichiometric ratios of interest were P/HAc, PHV/PHB, PHV/HAc, PHB/HAc and gly/HAc.

2.2.6 Estimation of intracellular Poly-P/VSS ratio

To assess the observed anaerobic stoichiometry and their potential relationship to the Poly-P/VSS ratio, the average estimated Poly-P/VSS ratio was determined for each anaerobic phase as the average between the initial and final Poly-P/VSS ratio of a specific anaerobic feeding phase. The Poly-P/VSS ratios in the different experiments were calculated on the basis of the ash content (Equation 1). To evaluate Equation 1, the initial Poly-P/VSS ratios of the biomass in the SBR reactors were also estimated using the steady-state mass balances (Equation 2). A comparison of data obtained with both equations is shown in Table 2. A detailed description of the equations and the assumptions underlying these equations is given in the supplementary materials (S1).

$$f_{P,pp,VSS} = ((ISS/TSS)/(1-ISS/TSS) - f_{ISSb,TSS} / (1- f_{ISSb,TSS})) * f_{P,ppASH} \qquad \text{(Eq. 1e)}$$

$$f_{P,ppVSS} = SRT/HRT *(T_{P,i} - T_{P,e})/ VSS - f_{P,bVSS} \qquad \text{(Eq 2g)}$$

where;

TSS: Total suspended solids

VSS: Volatile suspended solids

ISS: Inorganic suspended solids

$T_{P,i}$: Total phosphorus concentration in the influent

$T_{P,e}$: Total phosphorus concentration in the effluent

$f_{P,ppVSS}$: Ratio of Poly-P per VSS

$f_{P,bVSS}$: Ratio of non Poly-P phosphorus per VSS (0.023mgP/mgVSS)

$f_{ISSb,TSS}$: Ash content associated with active biomass

$f_{P,ppASH}$: P-content of Poly-P (0.31mgP/mgISS)

HRT: Hydraulic retention time

SRT: Solids retention time

2.3 Results

2.3.1 Enrichment of PAO cultures
All reactors (SBR-L, SBR-W and SBR-S) operated under different conditions for approximately 925, 120 and 442 days, respectively, showing stable biomass activities during the execution of this study (Figure S2.1, Supplementary data). Pseudo steady-state conditions were confirmed, prior to the characterization of the biomass activity and microbial composition of the biomass in each reactor, by the stable online pH and DO profiles together with the measured orthophosphate and MLSS data.

2.3.1.1 Anaerobic conversions in parent reactors
In all reactors the volatile fatty acids (VFA) were consumed within 30 min with active biomass specific VFA uptake rates of 150, 41 and 170 (C-mmol VFA/C-mol.h) in SBR-L, SBR-W and SBR-S, correspondingly (Figure 2.1). During the anaerobic VFA-uptake , a mixed PAO and GAO metabolism was observed in SBR-L and SBR-W while in the anaerobic phase of SBR-S a typical PAO metabolism was observed as can be seen from the average P/VFA ratios at the end of the anaerobic phase were 0.17 ± 0.04, 0.34 ± 0.02 and 0.61 ± 0.13 P-mol/C-mol VFA. For comparison of the anaerobic stoichiometry of SBR-L and SBR-W (operated at pH 7.0 and HAc as carbon source) and SBR-S (operated at pH 7.6, 75% with HAc and 25% HPr as carbon source), one additional batch test was conducted with the SBR-S biomass at pH 7.0 and HAc. In this anaerobic batch test, the P/HAc ratio was 0.64 P-mol/C-mol HAc.. Table 2.2 shows the anaerobic stochiometric parameters obtained from a characteristic cycle in all reactors, together with the anaerobic stoichiometric parameters obtained in selected previous studies with enriched PAO and GAO cultures. The relatively high gly/HAc, PHV/PHB and PHV/HAc ratios and low P/HAc, PHB/gly and PHV/gly ratios for the conversions in SBR-W and SBR-L indicate that a GAO metabolism was significantly involved in the anaerobic HAc uptake.

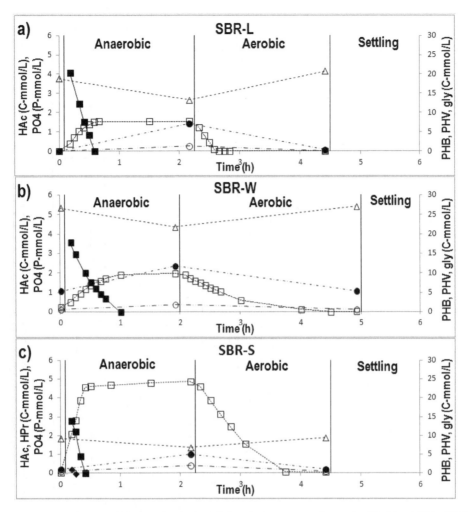

Figure 2.1 Measurements of a single cycles obtained during steady-state operation on day 890, 87 and 428 in SBR-L, SBR-W and SBR-S, respectively. Figure shows acetate (■), propionate (◆), PO₄ (□), PHV (○), PHB (●), and glycogen (Δ) concentrations for: (a) SBR-L, (b) SBR-W and (c) SBR-S reactor.

2.3.1.2 Aerobic conversions in parent reactors

In all systems, full P-removal (zero effluent P - all P limiting) was observed in the aerobic phase, indicating that the systems were cultivated under phosphate limiting conditions. During a characteristic cycle (Figure 2.1), ortho-phosphate was consumed in the aerobic phase at a specific P-uptake rate of 0.066 P-mol/(C-mol active biomass.h), 0.011 P-mol/(C-mol active biomass.h) and 0.090 P-mol/(C-mol.h) in SBR-L, SBR-W and SBR-S, respectively. The P-uptake rate determined at 20°C in SBR-L was lower than the rate from SBR-S. Similarly, the uptake rate determined at 10°C in SBR-W was lower than that reported by Brdjanovic *et al.* (1997) of 0.018 P-mol/(C-mol active biomass.h) at the same temperature. The VSS concentrations in SBR-L, SBR-W and SBR-S were 2300 ± 220 mg/L, 4000 ± 280 mg/L and 1674 ± 104 mg/L, respectively, resulting in MLVSS/MLSS ratios of 0.74 ± 0.02, 0.67 ± 0.05 and 0.63 ± 0.04. High ash contents are associated with high Poly-P contents. In spite of the full P-removal, from the anaerobic conversions, it was suspected that SBR-W and SBR-L contained GAO. This was also suggested by the lower P-uptake rates of these cultures.

31

Reference	Reactor	(Suspected) organisms (P)AO, (G)AO	SRT (d)	HRT (h)	MLVSS/MLSS (mg/mg)	Poly-P/MLVSS[d] (ash) (mg/mg)	Poly-P/MLVSS[e] (P-balance) (mg/mg)	GAO[f] % of EUB	PAO % of EUB	PAO I % of PAO	PAO II % of PAO	pH	Influent $[Ca^{2+}]$ (mg/L)	PHV/PHB (C-mol/C-mol)	PHV/HAc (C-mol/C-mol)	PHB/HAc (C-mol/C-mol)	P/HAc (P-mol/C-mol)	Gly/HAc (C-mol/C-mol)	Δ-PHV/PHB (C-mol/C-mol)	Δ-P/HAc (P-mol/C-mol)	(Δ-PHV/PHB)/(Δ-P/HAc) (P-mol/C-mol)	$q_{SA,ana}^{MAX}$ [C-mmol/(C-mol.h)]	$q_{P,aer}^{MAX}$ [P-mmol/(C-mol.h)]
This study	SBR-L	P	8	12	0.75	0.10	0.082	0	99±3	1±2	99±6	7 ±0.05	3.8	0.19	0.23	1.24	0.22	0.96	0.08	0.21	0.38	154	66
	B-L	P	NA	NA	0.93	0.01	NA	NA	NA	NA	NA	7 ± 0.1	3.8	0.27	0.32	1.19	0.01	0.98				80	NA
	SBR-W[c]	P	32	12	0.67	0.17	0.18	0.2±1	87±14	38±26	61±28	7 ± 0.1	3.8	0.21	0.29	1.36	0.37	1.11	0.11	0.45	0.45	41	11
	B-W	P	NA	NA	0.85	0.05	NA	NA	NA	NA	NA	7 ± 0.1	3.8	0.32	0.43	1.34	0.13	1.06	NA	NA	NA	62	NA
	SBR-S	P	8	12	0.58	0.22	0.21	0	99±8	98±9	0	7.6 ±0.1	3.8	0.36	NA	NA	NA	NA	0.26	0.62	0.43	NA	90
	B-S (high poly-P)	P	NA	NA	0.58	0.21	NA	NA	NA	NA	NA	7 ± 0.1	3.8	0.07	0.09	1.27	0.64	0.29				179	NA
	B-S	P	NA	NA	0.95	0.00	NA	NA	NA	NA	NA	7 ± 0.1	3.8	0.33	0.37	1.09	0.02	1.28				23	NA
Brdjanovic et al. (1997, 1998)[a]	SBR-D	P	8	12	0.68	NA	0.08	NA	NA	NA	NA	7 ± 0.1	3.8	0.09	0.06	0.71	0.45	0.4	0.09	0.22	0.40	NA	37,18[c]
	B-D	P	NA	NA	0.9	NA	NA	NA	NA	NA	NA	7 ± 0.1	3.8	0.18	0.25	1.39	0.23	0.58	NA	NA	NA	NA	NA
Zhou et al. (2008)	SBR	P	8	24	0.6	NA	0.17	0	80	NA	NA	7.0-8.0	1.3	0.06	0.07	1.18	0.62	0.46	0.37	0.54	0.68	NA	NA
	B	P	NA	NA	NA	NA	NA	0	80	NA	NA	7.5±0.01	1.3	0.37	0.46	1.24	0.06	1.03	NA	NA	NA	NA	NA
Acevedo et al. (2012)	SBR	P	8	12	0.45	NA	0.28	0	87	66	8	7.0-8.9	10	0.04	0.05	1.31	0.7	0.38	0.12	0.62	0.19	NA	NA
	SBR	P	8	12	0.92	NA	0.01	0	86	23	36	7.0-8.9	10	0.16	0.28	1.74	0.08	1.08	NA	NA	NA	NA	NA
Smolders et al. (1994, 1995)	SBR	P	8	12	0.69	NA	0.09	NA	NA	NA	NA	7 ± 0.05	3.8	0.1	0.12	1.21	0.5	0.5	NA	NA	NA	NA	NA
Tian et al. (2013)	SBR	P	16	12	NA	NA	NA	18	81	100	0	7 ± 0.1	3.8	0.1	0.13	1.31	0.56	0.55	NA	NA	NA	NA	NA
Zeng et al. (2003)	SBR	G	6.6	8	0.97	NA	0	NA	NA	NA	NA	7 ± 0.1	6.8	0.38	0.52	1.39	NA	1.20	NA	NA	NA	NA	NA
Lopez-Vazquez et al. (2007)[b]	SBR	G	10	12	0.9		0	75	20	NA	NA	7 ± 0.1	3.8	0.34	0.69	1.28	0.01	1.20	NA	NA	NA	NA	NA

Table 2.2 Stoichiometric and kinetic values of the anaerobic conversions observed in SBR-L, SBR-W and SBR-S in comparison to values reported in previous studies with enriched PAO and GAO cultures at 20°C (except data from SBR-W and Brdjanovic et al. (1997) which correspond to an operational temperature of 10°C).

Accumulibacter clades 'Type I and II performing kinetically different glycogen-accumulating organisms metabolisms for anaerobic substrate uptake

Note: B-L, B-W and B-D show the characteristics of the final cycle of the batch tests conducted with sludge from SBR-L, SBR-W, SBR-S and a batch test described by Brdjanovic *et al.* (1998), respectively. NA = not applicable, ND = not detected (with the used FISH probes).

a: Stoichiometric values calculated with data provided in Table 1 in Brdjanovic *et al.* (1998).

b: Stoichiometric values obtained from figures.

c: Experiments conducted at 10°C

d: Calculated with equation 1 in material and methods

e: Calculated with equation 2 in material and methods

f: GAO refers to '*Candidatus* Competibacter phosphatis' targeted by GAOMIX probe (mixture of probes GAOQ431 and GAOQ989) (Crocetti *et al.*, 2002)

2.3.1.3. Confirmation of truly anaerobic conditions and absence of chemical precipitation

Additional measurements were undertaken to confirm that the observed conversion processes were not influenced by the potential presence of any external electron acceptor in the anaerobic stage or affected by phosphate precipitation. The redox potential under anaerobic conditions was -220mV, while dissolved oxygen, nitrate or nitrite were not detected, indicating truly anaerobic conditions. The anaerobic release and aerobic uptake of potassium (K^+) magnesium (Mg^{2+}) and calcium (Ca^{2+}) in SBR-L and SBR-W showed no considerable differences between the anaerobic and aerobic K^+/PO_4-P and Mg^{2+}/PO_4-P ratios (SBR-L: release, 0.227 K^+-mol/P-mol, 0.302 Mg^{2+}-mol/P-mol; uptake 0.234 K^+-mol/P-mol, 0.304 Mg^{2+}-mol/P-mol; SBR-W: release, 0.255 K^+-mol/P-mol, 0.258 Mg^{2+}-mol/P-mol; uptake 0.287 K^+-mol/P-mol, 0.272 Mg^{2+}-mol/P-mol). Negligible differences in calcium concentrations were observed, indicating that Ca^{2+} did not play a significant role in biological or chemical phosphate removal. Based on these results, the potential precipitation of phosphorus was discarded.

2.3.1.4 Characterization of microbial population

FISH analysis demonstrated that all reactors were highly enriched with '*Candidatus* Accumulibacter phosphatis', as seen from the matching surface area coverage of the PAOmix and EUB338mix probes in the FISH images (Figures 2.2a, 2.2b and 2.22c or Figure S2.2, Supplementary data). For the cultures that exhibited a partial GAO metabolism, this was double confirmed by an additional FISH analysis, using probes specific for the *β-Proteobacteria* (subdivision to which most '*Candidatus* Accumulibacter phosphatis' belong) (Figure S2.2, Supplementary data). With the specific GAO mix probes applied in this study, '*Candidatus* Competibacter phosphatis' was not detected at all and the absence of GAO was also double confirmed in SBR-L and SBR-W by an additional FISH analyses with *γ-Proteobacteria* probes (the subdivision to which '*Candidatus* Competibacter phosphatis' belongs) (Figure S2.2. Supplementary data). *Defluviicoccus* was not considered as a potential GAO in SBR-L and SBR-W, the systems that exhibited a partial GAO metabolism, as these systems were operated with HAc as the sole carbon source, while *Defluviicoccus* is only competitive with *Accumulibacter* in acetate fed systems (Lanham et al., 2008, Lopez-Vazquez et al., 2009, Carvalheira et al., 2014). Finally, neisser staining tests showed that intracellular Poly-P was homogeneously distributed over the organisms (data not shown), supporting the observations in the FISH analysis that the sludge was dominated with PAO.

FISH quantification showed that PAO comprised around 99±3, 87±14 and 99±8% of the total bacterial population (with regard to EUB) in SBR-L, SBR-W and SBR-S, respectively. Further FISH analysis showed that PAO II was the dominant PAO population in SBR-L (corresponding to 99±6 % of total PAO), as shown in Figure 2.2d and PAO I in SBR-S (corresponding to 98±9 % of total PAO), as shown in Figure 2.2e. Meanwhile SBR-W contained a mixture of PAO I and II (61±28 and 38±26%, respectively) as shown in Figure 2.2f.

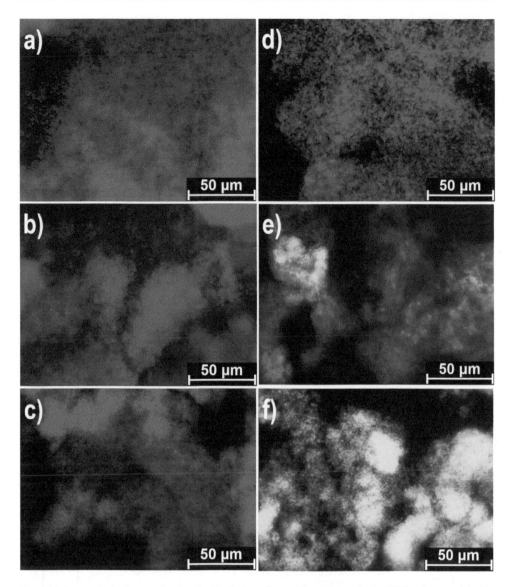

Figure 2.2 Representative images showing the distribution of bacterial populations in the EBPR cultures enriched in SBR-L (a and d), SBR-W (b and e) and SBR-S (c and f) by applying FISH. In Figures a, b and c; blue: EUB mix (Cy5); purple (superposition of blue and red): PAO mix (Cy3); and cyan green (superposition of blue and green): GAO mix (Fluos). In Figures d, e and f; blue: PAO mix (Cy5), purple (superposition of blue and red): PAO clade II (Cy3), and cyan green (superposition of blue and green): PAO clade I (Fluos).

Additional DGGE analysis of 16S rRNA gene fragments was performed for qualitative verification regarding the presence of specific GAO and PAO clades. For comparison purposes, an additional DGGE analysis was carried out on a stored sample from a previous study carried out at 10°C (Tian *et al.*, 2013) (herein labeled as SBR-C). The DNA derived patterns of the samples showed 10, 11, 6 and 4 bands for SBR-L, SBR-W, SBR-S and SBR-C, respectively (Figure S2.3, Supplementary data). The phylogenetic analysis did not detect any known *Competibacter* or *Defluviicoccus* population (known GAO) (Figure 2.3). Neither were γ-*Proteobacteria* (the subdivision to which '*Candidatus* Competibacter phosphatis' belongs) detected. In SBR-L, one dominant and several minor PAO species were observed, all closely related to '*Candidatus* Accumulibacter phosphatis' Types IIC/D. In SBR-W, a mixed population of two PAO clades, closely related to '*Candidatus* Accumulibacter phosphatis' Clade I and Clade IIA were detected. In SBR-S and SBR-C samples, only PAO species closely related to '*Candidatus* Accumulibacter phosphatis' Clade I were observed. The presence of other species belonging to α-*Proteobacteria*, β-*Proteobacteria*, δ-*Proteobacteria*, *Armatimonadetes* and *Bacteroidetes* were considered negligible since practically all cells in the biomass stained positive with the β-*Proteobacteria* and PAO mix FISH probes.

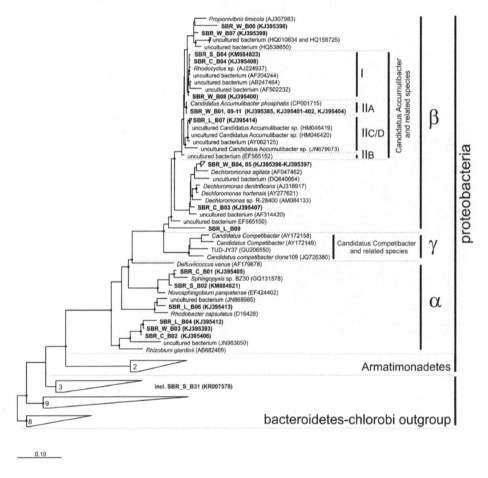

Figure 2.3 Phylogenetic analysis of partial 16S rRNA gene sequences obtained from DGGE bands from biomass samples collected in SBR-L, SBR-W, SBR-S and SBR-C. In total, 522 positions were used for calculation. Bootstrap was performed and values ≥70% are indicated by a •. The scale bar below indicates a 10% sequence difference.

2.3.2 Anaerobic HAc uptake by PAO depleted in Poly-P content

2.3.2.1 Effect of Poly-P depletion on HAc uptake

Batch tests were performed at pH 7.5 to deplete the Poly-P content of the biomass and after depletion of Poly-P at pH 7.0 to assess the ability of the enriched cultures to take up acetate under conditions where the Poly-P contents were stoichiometrically limiting the VFA uptake and to compare the stoichiometric and kinetic values to those reported in literature (most of them at pH 7.0) (Figure 2.4). After the Poly-P depletion cycles (three consecutive cycles for SBR-L and SBR-W, and five for SBR-S), the average estimated Poly-P/VSS ratios reached practically zero in the SBR-L and SBR-S cultures, but in SBR-W Poly-P was not fully depleted (0.04mgP/mgVSS). Likely, this was caused by the slightly higher Poly-P content of the sludge and the almost double VSS concentration in SBR-W with regard to other reactors. As the same HAc-feed was applied in each poly-P depletion batch test, the HAc-feed/VSS was about half the HAc-feed/VSS of SBR-L and SBR-S. Consequently, more cycles were needed to reach the poly-P depleted stage. In all batch tests at pH 7.0, HAc uptake was observed at P/HAc ratios of 0.01 and 0.13 and 0.02 P-mol/C-mol for biomass from SBR-L, SBR-W and SBR-S, respectively. At pH 7.5 the HAc uptake seemed to become limited by the Poly-P content (Table 2.2). Interestingly and in particular, the specific HAc uptake rate of PAO I in SBR-S decreased 7 times (which initially contained a high Poly-P/VSS ratio) while the rate of SBR-L enriched with PAO II decreased only 2 times (Table 2.2).

37

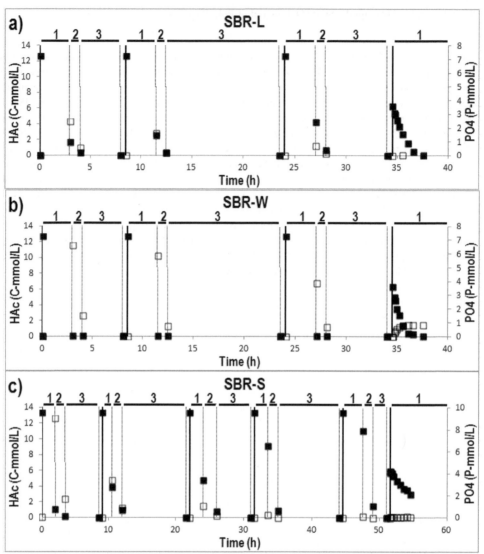

Figure 2.4 Results obtained from the batch tests executed at 20 oC with biomass from: (a) SBR-L, (b) SBR-W, and (c) SBR-S with: Acetate (■) and PO₄ (□). Bold line: start of feeding phase. Dashed lines: start of washing, aerobic or settling phases. Numbers along the horizontal bar on top of figures represent the: (1) anaerobic phase, (2) washing phase, and (3) aerobic phase.

2.3.2.2 Effect of Poly-P depletion on the anaerobic stoichiometry

In all reactors, the stoichiometry of the anaerobic conversions showed that a decrease in P-release coincided with a higher PHV/PHB ratio and, in SBR-L and SBR-S, with a higher gly/HAc ratio (Table 2.2). This indicates that the decreased ATP generation from Poly-P was compensated by extra ATP generation from glycolysis. During the depletion of Poly-P in the consecutive feeding and washing steps at pH 7.5, a gradual change in the P/HAc ratio was observed, which correlated with the average estimated Poly-P/VSS ratio of the biomass (Figure 2.5b). Interestingly, it seems that at the higher pH a minimal P/HAc ratio is observed whereas at pH 7.0 the P-release seems to be not essential for acetate uptake (Figure 2.5a).

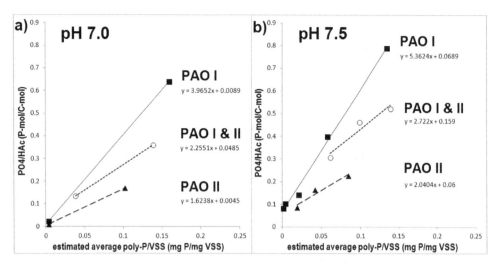

Figure 2.5 Stoichiometric P released/HAc consumed values from the Poly-P depletion tests observed at: (a) pH 7.0 and (b) pH 7.5. (▲) P released/HAc consumed in SBR-L; (○) P released/HAc consumed in SBR-W; and, (■) P-released/HAc consumed in SBR-S. The estimated average Poly-P/VSS ratio was determined for each anaerobic phase as the average between the initial and final estimated Poly-P/VSS ratio on the basis of the ash content, using equation 1.

2.4 Discussion

2.4.1 PAO clades exhibiting a GAO metabolism during long-term performance

In this study, highly enriched cultures of PAO were cultivated (>87% enrichment), whereas no GAO ('*Candidatus* Competibacter phosphatis') were detected (based on both FISH and DGGE analyzes). A PAO II culture was enriched in SBR-L (99±6% with regard to PAOmix), a PAO I culture in SBR-S (98±9 % of PAOmix), while in SBR-W a mixture of PAO I and II was obtained (61±28 and 38±26%, with regard to PAOmix). Previously, Tian *et al.* (2013) enriched a PAO I culture in SBR-C at 10°C with a SRT of 16 days (6 days aerobic SRT). The PAO I cultures cultivated in SBR-S (this study) and SBR-C (Tian et al., 2013) exhibited a typical PAO metabolism. In contrast, the PAO II culture and mixed PAO I and II culture enriched in this study exhibited a partial GAO metabolism in the anaerobic stage under conditions where Poly-P was not stoichiometrically limiting the VFA uptake, when corresponding operating conditions were truly anaerobic and any potential chemical phosphorus precipitation was discarded. Moreover, the pH and Ca^{2+} concentration in SBR-L and SBR-W were similar to those from SBR-S and previously reported values (Table 2.2) (Smolders *et al.*, 1994; Brdjanovic *et al.*, 1998; Tian *et al.*, 2013). Therefore, their influence (known to affect the EBPR stoichiometry) did not play a role when comparing these and previous studies. These findings suggest that PAO II can exhibit a partial GAO metabolism (i.e. Glycolysis can supply a surplus of reducing equivalents used to convert HAc to PHV) during long-term performance under conditions where phosphate is limiting (all phosphate was taken up by the end of the aerobic phase), but where the availability of Poly-P is not stoichiometrically limiting the VFA uptake, while PAO I exhibits the typical PAO metabolism (i.e. Glycolysis only used to supply redox equivalents for conversion of HAc to PHB).

2.4.2 Effect of P-content on stoichiometry

P-depletion tests showed that both PAO clade I and II are able to gradually shift their metabolism towards a GAO metabolism when the Poly-P content decreases. This observation is in agreement with the findings from Brdjanovic *et al.* (1998), Zhou et al. (2008), Erdal et al. (2008) and Acevedo *et al.* (2012), who observed a shift from a PAO metabolism towards a (partial) GAO metabolism when the Poly-P content decreased. In the study by Acevedo et al. (2012), a significant change in the PAO I and II fractions were observed (decrease of PAO I fraction from 54 to 32 % and increase of PAO II fraction from 27% to 48%) during a period of 8 cycles (0.25 x SRT) when the poly-P content of the sludge was decreased in a mixed PAO I and II culture, cultivated with a SRT of 8 days (32 cycles). In the present study it was expected that such a population change would not be possible for the following reasons. Unlike the study by Acevedo et al. (2012), in SBR-L and SBR-S (also cultivated with a SRT of 8 days - 32 cycles), the PAO I (SBR-S) and PAO II (SBR-L) cultures were highly enriched with the specific PAO clades. In SBR-S PAO II and GAO were below detection level and in SBR-L, GAO were below detection level and PAO I were just 1%). Therefore a rapid shift in the population is not possible in the duration of the experiment (3 to 4 cycles with double HAc load, equivalent to 6 to 8 cycles). On the other hand SBR-W, did contain a mixed PAO I and II culture, similar to the studies of Acevedo et al., 2012), but considering the much longer SRT of 32 days (128 cycles) in this study for the mixed PAOI-PAOII culture (SBR-W) and the short duration of the experiment 3 cycles with double HAc-load (equivalent to 6 cycles = 0.05 x SRT)), a significant change in the microbial community was not considered likely, even if one PAO clade would have been fully inhibited and another clade fully active.

Although the changes in the PHV/PHB and gly/HAc ratios observed for SBR-L (PAO II) and SBR-W (mixed PAO I and II) biomasses were not as pronounced as the observed changes for the biomass from SBR-S (PAO I) and the study of Zhou et al. (2008) and Acevedo et al. (2008), the changes were still significant. The biomasses from SBR-L and SBR-W were already partially using a GAO metabolism under conditions where Poly-P was not stoichiometrically the VFA uptake and therefore the difference in the PHV/PHB ratio between the Poly-P non-depleted and Poly-P depleted biomass is not as pronounced as the difference observed in SBR-S and previous studies (Zhou et al., 2008; Acevedo at al., 2012), in which the biomass performed a PAO metabolism under Poly-P sufficient conditions with a high P/HAc ratio of 0.6 and 0.7, respectively. This is further supported by the increase in PHV/PHB ratio proportional to the decrease in P/HAc in SBR-L and SBR-W (Table 2.2), which is for SBR-L and SBR-W in the same range as observed in SBR-S and in previous studies. In addition, the final PHV/PHB values obtained under Poly-P depleted conditions are in the range of those observed for the GAO-like metabolism by PAO in the studies by Zhou et al. (2008) and Acevedo et al. (2012) as well as the GAO PHV/PHB (0.38) ratio reported by Lopez-Vazquez et al. (2007), indicating that under Poly-P depleted conditions the biomass of SBR-L and SBR-W performed a GAO metabolism.

2.4.3 HAc uptake under Poly-P depleted conditions

Batch tests showed that PAO I and PAO II have significant differences in their HAc-uptake rates under Poly-P depleting conditions. The active biomass specific HAc-uptake rate under Poly-P depleted conditions in comparison to the rates under steady-state operation of PAO I and PAO II decreased 7 times and 2 times, respectively. This finding is in line with the observations of Tian *et al.* (2013) where, during long-term experiments, a decrease in the influent phosphate concentration led to limited HAc uptake by a PAO I culture and eventually to EBPR deterioration.

Furthermore, the HAc-uptake rate of PAO II was four times higher than that of PAO I under Poly-P depleted conditions, indicating that when Poly-P is limiting the VFA-uptake, PAO II has a strong competitive advantage over PAO I. This is accordance to the findings of Acevedo et al. (2012) who observed a shift in the microbial population from a highly enriched PAO I to a mixed PAO I and II culture when the Poly-P content of the biomass was significantly reduced.

Differences in the ability to take up HAc in the absence of Poly-P among different PAO clades may be dependent on the glycolysis pathway (Lopez-Vazquez *et al.*, 2008). It cannot be discarded that different PAO clades could use different pathways as both the Entner-Doudoroff (ED) pathway and the Embden-

Meyerhoff-Parnas (EM) pathway have been proposed for glycogen degradation in enriched PAO cultures based on 13C-NMR and genomic studies (Maurer *et al.*, 1997; Pereira *et al.*, 1996; Hesselman *et al.*, 2000; Martin *et al.*, 2006). Further research on highly enriched PAO I and PAO II cultures could contribute to unveil whether they use different glycolytic pathways.

2.4.4 Intrinsic physiological differences between Accumulibacter clades

The profiles of the P/HAc ratios versus the average estimated Poly-P/VSS ratio showed that each culture had a different P/HAc stoichiometry even at similar Poly-P/VSS ratios (Figure 2.5). In addition, PAO I and II showed different uptake rates as the intracellular Poly-P content decreased (Table 2.2). These findings support the hypothesis that very significant intrinsic differences exist in the anaerobic stoichiometry of PAO I and II, with PAO I showing higher P/HAc ratios than PAO II and that the differences in metabolism observed during long-term operation are not only caused by differences in the estimated Poly-P/VSS ratio. These findings can help to explain the broad range of stoichiometric values observed for enriched EBPR biomass with similar P-content in previous studies (Cech and Hartman, 1993; Liu et al., 1997; Kisoglu et al., 2000; Sudiana et al., 1999; Schuler and Jenkins, 2003), and highlight the need to perform a better evaluation of the micro-diversity of PAO populations in experimental systems.

A schematic representation of the proposed source of NADH and ATP for the uptake of VFA by GAO, PAO II and PAO I is shown in Figure 2.6. PAO I uses glycolysis for the supply of NADH for reduction of HAc into PHB. This generates ATP but additional ATP needed for the sequestration of HAc into PHA is derived from Poly-P conversion. GAO make use of the glycolysis to deliver 100% of the ATP and NADH. They produce a surplus of NADH which is balanced by converting a fraction of acetyl-CoA into propionyl-CoA, resulting in PHV formation. PAO II are more flexible by using a mixture of the PAO I and GAO metabolism.

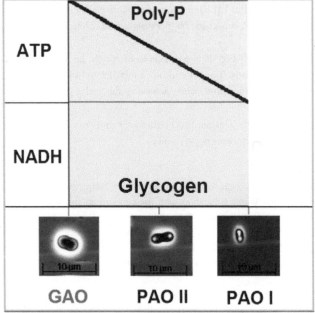

Figure 2.6 Schematic representation of the source of NADH and ATP for anaerobic uptake of VFA for GAO, PAO I and PAO II under Poly-P non-depleted conditions. Phase contrast images of PAO I and II were obtained in this study and the phase contrast image of GAO was obtained from an unpublished study.

2.4.5 Implications

This study provides evidence that the dominant PAO clade affects the anaerobic stoichiometry of enriched EBPR cultures under similar operating conditions (e.g. varying from 0.23 to 0.52 P-mol/C-mol at similar Poly-P/VSS ratios, pH, carbon source and calcium concentration). In systems for biological phosphorus recovery or combined biological and chemical P-removal systems (such as the PhoStrip or BCFS process)(Brdjanovic et al., 2000; Barat et al., 2006b), the dominant PAO clade likely influences the P-release efficiency (P-release/HAc-feed) affecting the plant performance (e.g. PAO I have a P-release activity 2.3 times higher than that of PAO II). Nevertheless, since the highest P-release occurs close to Poly-P saturation (when the PAO have reached their maximum storage capacity), adjusting the VFA load, and thereby the PAO biomass formation, may help to improve the P-release process efficiency. The influent phosphorus concentration and periodical P-content depletion can be two factors affecting the occurrence of PAO I and II. Likely, PAO I will dominate at higher P-influent concentrations (e.g. higher than 15 mg/L), while PAO II could lead under lower influent P concentrations and in biomass exposed to periodic Poly-P depleting conditions due to their higher metabolic flexibility as suggested by Acevedo et al. (2012). Nevertheless, the role of other factors (such as pH, VFA composition, temperature and dissolved oxygen,) cannot be discarded. These findings also suggest that the niche differentiation between PAO clades needs to be taken into account to improve the description and reliability of activated sludge models and sludge characterization methods.

2.5 Conclusions

In the present study, it was demonstrated that:

- When the Poly-P content decreases at short-term, both '*Candidatus* Accumulibacter phosphatis' Type I and II can shift their metabolism from a PAO or mixed PAO-GAO metabolism to a GAO metabolism and have the ability to solely rely on glycogen as energy source for HAc uptake.
- Under Poly-P depleted conditions, the kinetic rates of PAO II are four times higher than those of PAO I, suggesting that PAO II has a strong competitive advantage under low P/VFA conditions or when the influent P concentration fluctuates, meanwhile PAO I may dominate at higher influent P/VFA ratios.
- Under conditions where Poly-P is not stoichiometrically limiting the VFA-uptake, '*Candidatus* Accumulibacter phosphatis' Type II can use a mixed PAO-GAO metabolism in the anaerobic stage during a long-term period, while achieving full ortho-phosphate removal under aerobic conditions.
- The metabolic differences between PAO clades are factors that have contributed to the variations in PAO and GAO metabolisms in past studies.

2.6 Acknowledgement

This study was carried out as part of the saline project (http://www.salinesanitation.info) led by UNESCO-IHE Institute for Water Education with the consortium partners KWR Watercycle Research Institute, Delft University of Technology, University of Cape Town, The Hong Kong University of Science and Technology, Polytechnic University José Antonio Echeverría and Birzeit University. The saline project is financed by UNESCO-IHE internal research fund and with a special generous contribution from Professor George Ekama from University of Cape Town.

2.7 References

Acevedo, B., Oehmen, A., Carvalho, G., Seco, A., Borras, L., and Barat, R. (2012) Metabolic shift of polyphosphate-accumulating organisms with different levels of poly-phosphate storage. Water Res 46: 1889-1900

Amann, R.I., Binder, B.J., Olson, R.J., Chisholm, S.W., Devereux, R., and Stahl, D.A. (1990) Combination of 16S rRNA-targeted oligonucleotide probes with flow cytometry for analyzing mixed microbial populations. Appl Environ Microbiol 56: 1919-1925

A.P.H.A. (1995) Standard Methods for the Examination of Water and Waste Water, 17th Edn. Washington DC: American Public Health Association.

Barat, R., Montoya, T., Borras, L., Seco, A. and Ferrer, J. (2006) Calcium Effect on Enhanced Biological Phosphorus Removal. Water Sci Technol 53(12): 29-37.

Barat, R., and van Loosdrecht, M.C.M. (2006b) Potential phosphorus recovery in a WWTP with the BCFS process: Interactions with the biological process. Water Res 40(19): 3507-3516.

Barat, R., Montoya, T., Borras, L., Ferrer, J., and Seco, A. (2008) Interactions between calcium precipitation and the polyphosphate-accumulating bacteria metabolism. Water Res 42(13): 3415-3424

Barnard, J.L. (1975) Biological nutrient removal without addition of chemicals. Water Res 9(5-6): 458-490

Bassin, J.P., Pronk, M., Muyzer, G., Kleerebezem, R., Dezotti, M., and van Loosdrecht, M.C.M. (2011) Effect of Elevated Salt Concentrations on the Aerobic Granular Sludge Process: Linking Microbial Activity with Microbial community Structure. Appl Environ Microbiol 77 (22): 7942-7953.

Brdjanovic, D., van Loosdrecht, M.C.M., Hooijmans, C.M., Alaerts, and G.J., Heijnen J.J., (1997) Temperature Effects on Physiology of Biological Phosphorus Removal. J Environ Eng-ASCE 123(2): 144-154

Brdjanovic, D., van Loosdrecht, M.C.M., Hooijmans, C.M., Mino, T., Alaerts G.J., and Heijnen, J.J., (1998) Effect of polyphosphate limitation on the anaerobic metabolism of phosphorus-accumulating microorganisms. Appl. Microbiol Biotechnol 50: 273-276

Brdjanovic, D., van Loosdrecht, M.C.M., Versteeg, P., Hooijmans, C.M., Alaerts, G.J. and Heijnen, J.J. (2000) Modeling COD, N and P removal in a full-scale sstp Haarlem Waarderpolder. Water Res 34(3): 846-858

Carvalho, G., Lemos, P.C., Oehmen, A., and Reis, M.A.M. (2007) Denitrifying phosphorus removal: linking the process performance with the microbial community structure. Water Res 41(19): 4383, 4396

Cech, J.S., and Hartman, P., (1993) Glucose-Induced between Polyphosphate and Polysaccharide Accumulating Accumulating Bacteria in Ehanced Biological Phosphorus Removal. Water Res 20: 1511

Comeau, Y., Hall, K.J., Hancock, R.E.W., and Oldham, W.K. (1986) Biochemical-model for enhanced biological phosphorus removal. Water Res 20 (12): 1511–1521.

Crocetti, G. R., Hugenholtz, P., Bond, P.L., Schuler, A., Keller J., Jenkins, D., and Blackall, L.L., (2000) Identification of polyphosphate-accumulating organisms and design of 16S rRNA-directed probes for their detection and quantitation. Appl Environ Microbiol 66: 1175-1182

Crocetti, G.R., Banfield, J.F., Keller, J., Bond, P.L., and Blackall, L.L. (2002) Glycogen accumulating organisms in laboratory-scale and full-scale wastewater treatment processes. Microbiol 148: 3353-3364.

Daims, H., Bruhl, A., Amann, R., Amann, R., Schleifer, K.H., and Wagner, M. (1999) The domain-specific probe EUB338 is insufficient for the detection of all bacteria: development and evaluation of a more comprehensive probe set. Syst Appl Microbiol 22: 345-352

Ekama, G. A., & Wentzel, M. C. (2004). A predictive model for the reactor inorganic suspended solids concentration in activated sludge systems. Water research, 38(19), 4093-4106.

Erdal, U.G., Erdal, Z.K., Daigger, G.T., and Randall, C.W. (2008) Is it PAO-GAO competition or metabolic shift in EBPR system? Evidence form an experimental study. Water Sci Technol 58(6): 1329-1334

Flowers, J.J., He, S., Yilmaz, S., Noguera, D.R., and McMahon, K.D. (2009) Denitrification capabilities of two biological phosphorus removal sludges dominated by different 'Candidatus Accumulibacter' clades. Environ Microbiol Rep 1(6): 583-588

Hesselman, R.P.X., von Rummell, R., Resnick, S.M., Hany, R., and Zehnder, A.J.B. (2000) Anaerobic Metabolism of Bacteria Performing Enhanced Biological Phosphate Removal. Water Res 34: 3487

Kisoglu, Z., Erdal, U., and Randall, C.W. (2000) The Effect of COD/TP ratio on Intracellular Storage Materials, System performance and Kinetic Parameters in a BNR system. Proceedings of the 73rd Annual Water Environment Federation Technical Exposition and Conference.

Kong, Y.H., Beer, M., Rees, G.N., and Seviour, R.J. (2002) Functional analysis of microbial communities in aerobic-anaerobic sequencing batch reactors fed with different phosphorus/carbon)P/C) ratios. Microbiology 148: 2299-2307

Liu, W.T., Nakamura, K., Matsuo, T., and Mino, T., (1997) Internal Energy-Based Competition Between Polyphosphate- and Glycogen-Accumulating Bacteria In Biological Phosphorus Removal Reactors-Effect of P/C Feeding Ratio. Water Res 31(6): 1430-1438

Lopez-Vazquez, C.M., Hooijmans, C.M., Brdjanovic, D., Gijzen, H.J., and van Loosdrecht, M.C.M., (2007). A Practical Method for the Quantification of Phosphorus- and Glycogen-Accumulating Organisms Populations in Activated Sludge Systems. Water Environ Res 79(13): 2487-2498.

Lopez-Vazquez, C.M., Brdjanovic, D., and van Loosdrecht, M.C.M. (2008) Comment on " Could polyphosphate-accumulating organisms (PAO) be glyccogen-accumulating organisms (GAO)?" by Zhou, Y., Pijuan , M., Zeng, R., Lu, Huabing and Yuan Z. Water Res. (2008) doi:10.1016/jwaterres.2008.01.003. Water Res 42: 3561-3562

Manz, W., Amann, R., Ludwig, W., Wagner, M., and Schleifer, K.H., (1992) Phylogenetic oligodeoxynucleotide probes for the major subclasses of Proteobacteria: problems and solutions. Syst Appl Micobiol 15: 593-600

Martín, H.G., Ivanova, N., Kunin, V., Warnecke, F., Barry, K.W., McHardy, et al., (2006) Metagenomic analysis of two enhanced biological phosphorus removal (EBPR) sludge communities. *Nat Biotechnol* **24:** 1263 - 1269

Maurer, M., Gujer, W., Hany, R., and Bachmann, S. (1997) Intracellular Carbon Flow in Phosphorus Accumulating Organisms from Activated Sludge Systems. Water Res 31(4): 907-917

Metcalf and Eddy, Inc. (2003) Wastewater Engineering - Treatment and Reuse, 4th Edn. New York, USA: Mc Graw Hill.

Mino, T., Arun, V., Tsuzuki, Y., and Matsuo, T., (1987) Effect of phosphorus accumulation on acetate metabolism in the biological phosphorus removal process. In Biological Phosphate Removal from Wastewaters, Advances in Water Pollution Control. Ramadori, R. (Ed). Oxford: Pergamon Press, pp. 27-38

Mino, T., Van Loosdrecht, M. C. M., & Heijnen, J. J. (1998). Microbiology and biochemistry of the enhanced biological phosphate removal process. Water research, 32(11), 3193-3207.

Oehmen, A., Lemos, P.C., Carvalho, G., Yuan, Z.G., Keller, J., Blackall, L.L., and Reis, M.A.M. (2010) Assessing the abundance and activity of denitrifying polyphosphate accumulating organisms through molecular and chemical techniques. Water Sci Technol 61(8): 2061-2068

Pereira, H., Lemos, P.C., Reis, M.A.M., Crespo, J.P.S.G. Carrondo, M.J.T., and Santos H. (1996) Model for Carbon Metabolism in Biological Phosphorus Removal Processes Based on In Vivo C13-NMR Labelling Experiments. Water Res 30: 2128

Schuler, A.J., and Jenkins, D. (2003) Enhanced Biological Phosphorus Removal from Wastewater by Biomass with Different Phosphorus Contents, Part 1: Experimental Results and Comparison with Metabolic Models. Water Environ Res 75(6): 485-498

Slater, F.R., Johnson, C.R., Blackall, L.L., Beiko, R.G., and Bond P.L. (2010) Monitoring associations between clade-level variation, overall community structure and ecosystem function in enhanced biological phosphorus removal (EBPR) systems using terminal-restriction fragment length polymorphism (T-RFLP). Water Res 44(17): 4908-4923

Smolders, G.J.F., van der Meij, J., van Loosdrecht, M.C.M., and Heijnen, J.J. (1994) Model of the Anaerobic Metabolism of the Biological Phosphorus Removal Process: Stoichiometry and pH influence. Biotechnol Bioeng (43): 461-470

Smolders, G.J.F., van Loosdrecht, M.C.M., and Heijnen, J.J. (1995) A Metabolic Model For the Biological Phosphorus Removal Process. Water Sci Technol (31): 79-97

Sudiana, I, Mino, T., Satoh, H., Nakamura, K., and Matsuo, T. (1999) Metabolism of Enhanced Biological Phosphorus Removal and Non-Enhanced Biological Phosphorus Removal Sludge with Acetate and Glucose as Carbon Source. Water Sci Technol (39): 29

Tian, W.D., Lopez-Vazquez, Li, W.G., C.M., Brdjanovic, Van Loosdrecht, M.C.M. (2013) Occurrence of PAOI in a low temperature EBPR system. Chemosphere 92: 1314-1320

Welles, L, Lopez-Vazquez, C.M., Hooijmans C.M., Van Loosdrecht, M.C.M., and Brdjanovic, D., (2014) Impact of salinity on the anaerobic metabolism of phosphate-accumulating organisms (PAO) and glycogen-accumulating organisms (GAO). Appl Microbiol Biotechnol 98(12): 7609-7622

Wentzel, M.C., Dold, P.L., Ekama, G.A., and Marais, G.v.R. (1985) Kinetics of biological phosphorus release. Water Sci Technol 17: 57-71

Wentzel, M.C., Dold, P.L., Loewenthal., R.E., Ekama, G.A., and Marais, G.v.R. (1987) Experiments towards Establishing the Kinetics of Biological Excess Phosphorus Removal. In Advances in Water Pollution Control: Biological Phosphate Removal from Wastewaters. Ramadori, R. (Ed). Oxford: Pergamon Press, pp. 79-91

Wentzel, M. C. (1988). Biological excess phosphorus removal in activated sludge systems.

Wentzel, M. C., Ekama, G. A., Loewenthal, R. E., Dold, P. L., & Marais, G. (1989a). Enhanced polyphosphate organism cultures in activated sludge systems. Part II: Experimental behaviour. Water S. A., 15(2), 71-88.

Wentzel, M. C., Dold, P. L., Ekama, G. A., & Marais, G. (1989b). Enhanced polyphosphate organism cultures in activated sludge systems. Part III: Kinetic model. Water S. A., 15(2), 89-102.Winkler, M. K., Bassin, J. P., Kleerebezem, R., De Bruin, L. M. M., Van den Brand, T. P. H., & Van Loosdrecht, M. C. M. (2011). Selective sludge removal in a segregated aerobic granular biomass system as a strategy to control PAO–GAO competition at high temperatures. Water research, 45(11), 3291-3299.

Zeng, R.J., Van Loosdrecht, M.C.M., Yuan, Z., and Keller, J., (2003). Metabolic model for glycogen-accumulating organisms in anaerobic/aerobic activated sludge systems. Biotechnol Bioeng 81(1): 92-105.

Zhou, Y., Pijuan, M., Zeng, R.J., Lu, H., and Yuan, Z. (2008) Could polyphosphate-accumulating organisms (PAO) be glyccogen-accumulating organisms (GAO)? Water Res 42: 2361-2368

2.8 Appendix 2A

Detailed description of intracellular Poly-P/VSS ratio estimation

The anaerobic stoichiometry was considered to be affected by the Poly-P/VSS ratio, but this ratio decreases during the HAc uptake mainly due to P-release. Consequently, the net HAc-uptake per VSS mass affects the Poly-P/VSS ratio directly and thereby indirectly the stoichiometry as well (Hesselman et al., 2000). To avoid differences in the stoichiometry due to differences in HAc-uptake per VSS mass, the average estimated Poly-P/VSS ratio was determined for each anaerobic phase as the average between the initial and final Poly-P/VSS ratio in a specific anaerobic feeding phase. The Poly-P/VSS ratios in the different experiments were calculated on the basis of the ash content (Equation 1). To evaluate Equation 1, the initial Poly-P/VSS ratios of the biomass in the SBR reactors were also estimated using the steady-state mass balances (Equation 2). A comparison of data obtained with both equations is shown in Table 2. Equation 1 was developed based on the ash content of the biomass assuming (i) an ash content associated with the active biomass (ISS_b) of 0.025 mg ISS/mgTSS (as observed in this study after Poly-P depletion); and, (ii) a Poly-P composition of $(PO_3)_3MgK$ with a P-content ($f_{P,ppASH}$) of 0.31 mgP/mgISS.Equation 2, derived from the steady-state mass balance of phosphorus, was drawn assuming (i) that the ratio of non Poly-P phosphorus per VSS ($f_{P,bVSS}$) is equal to the P-content of non-EBPR biomass which is 0.023 mgP/mgVSS (Metcalf and Eddy, 2003); and, (ii) a negligible concentration of solids in the effluent. To formulate both equations it was assumed that no chemical precipitation took place

$ISS_{pp} = ISS - ISS_b$ (Eq. 1a)

$ISS_b = f_{ISSb,TSS,} / (1- f_{ISSb,TSS}) * VSS$ (Eq. 1b)

$ISS_{pp} = ISS - f_{ISSb,TSS,} / (1- f_{ISSb,TSS}) * VSS$ (Eq. 1c)

$$f_{P,pp,VSS} = ISS_{pp} * f_{P,ppASH} / VSS \qquad \text{(Eq 1d)}$$

$$f_{P,pp,VSS} = ((ISS/TSS)/(1\text{-}ISS/TSS) - f_{ISSb,TSS} / (1\text{-} f_{ISSb,TSS})) * f_{P,ppASH} \qquad \text{(Eq. 1e)}$$

$$f_{P,VSS} = f_{P,ppVSS} + f_{P,bVSS} \qquad \text{(Eq 2a)}$$

$$f_{P,VSS} = f_{P,TSS} * TSS/VSS \qquad \text{(Eq 2b)}$$

$$d\,(T_{P,e}*V_p)/dt = Q_i * (T_{P,i} - T_{P,e}) - Q_w * TSS * f_{P,TSS} = 0 \qquad \text{(Eq 2c)}$$

$$Q_i / V_p * (T_{P,i} - T_{P,e}) - Q_w / V_p * TSS * f_{P,TSS} = 0 \qquad \text{(Eq 2d)}$$

$$1/HRT * (T_{P,i} - T_{P,e}) - 1/SRT * TSS * f_{P,TSS} = 0 \qquad \text{(Eq 2e)}$$

$$f_{P,TSS} = SRT/HRT * (T_{P,i} - T_{P,e})/ TSS \qquad \text{(Eq 2f)}$$

$$f_{P,ppVSS} = SRT/HRT *(T_{P,i} - T_{P,e})/ VSS - f_{P,bVSS} \qquad \text{(Eq 2g)}$$

where;

TSS: Total suspended solids

VSS: Volatile suspended solids

ISS: Inorganic suspended solids

ISS_b: Inorganic suspended solids associated with active biomass

ISS_{pp}: Inorganic suspended solids associated with Poly-P

$T_{P,i}$: Total phosphorus concentration in the influent

$T_{P,e}$: Total phosphorus concentration in the effluent

$f_{P,TSS}$: Ratio of total P per TSS

$f_{P,ppVSS}$: Ratio of Poly-P per VSS

$f_{P,bVSS}$: Ratio of non Poly-P phosphorus per VSS

$f_{P,VSS}$: Ratio of total P per VSS

$f_{ISSb,TSS}$: Ash content associated with active biomass

$f_{P,ppASH}$: P-content of Poly-P

V_p: Working volume of reactor

Q_i: Influent flow rate

Q_w: Wastage of biomass flow rate

HRT: Hydraulic retention time

SRT: Solids retention time

Although the ash content associated with active biomass (under Poly-P depleted conditions) obtained in this study was comparable with the ash content obtained in the GAO or mixed PAO-GAO cultures enriched under similar conditions with similar medium composition, the ash content values were lower than the values reported in some past studies. In a study by Ekama et al. (2004), a model was developed and calibrated, using data from 22 different studies on 30 aerobic and anoxic–aerobic nitrification–denitrification systems and 18 anaerobic–anoxic–aerobic ND biological excess P removal systems, variously fed artificial and real wastewater, and from 3 to 20 days sludge age. This study indicated that the ISS/TSS ratio associated to OHO biomass was 0.13. The significant difference in the ISS/TSS ratios observed in this study and those determined in the study of Ekama et al. (2004), maybe caused by differences in the concentrations of the minerals (in some cases 4x higher Ca^{2+} concentrations) and operational pH values (values up to 7.74) (Wentzel et al., 1989a). This suggests, that the equations provided in this study may only be used in for hihgly enriched PAO and GAO cultures that are cultivated under similar conditions.

On the basis of the measured release of Mg^{2+} and K^+, the Poly-P content of the biomass in this study was $(PO_3)_3Mg_{0.91}K_{0.69}$. However, as the positive charge balance covered only 83 to 84% of the negative charges from the phosphate units in Poly-P, it was assumed that in the bulk of poly-P inclusions contained more K and Mg resulting in the composition $(PO_3)_3MgK$.

2.9 Supplementary data

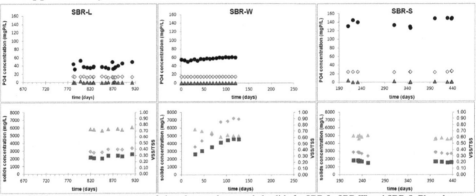

Figure S2.1 Long-term measurements of phosphate and suspended solids for SBR-L, SBR-W and SBR-S. Phosphate concentrations: at the end of the anaerobic phase (●), influent (◆), at the end of the aerobic phase (▲). Solids concentration: TSS (◆), VSS (■), VSS/TSS (▲).

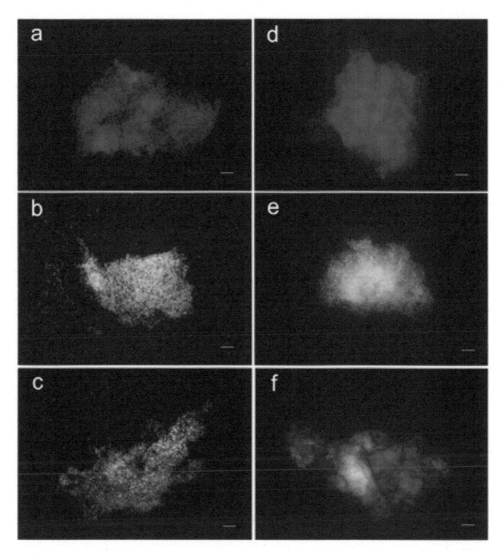

Figure S2.2 Distribution of bacterial polulation in the enriched cultures of polyphosphate-accumulating organisms in SBR-L (a, b and c) and SBR-W (d, e and f) by applying fluorescence in situ hybridization microscopy. Figures a and d, blue: Eubacteria (Cy5), purple (superposition of blue and red): Beta-proteobacteria (Cy3), and cyan green (superposition of blue and green): Gammaproteobacteria (Fluos). Figure b and e, blue: Eubacteria, purple (superposition of blue and red): Candidatus Competibacter phosphatis (Cy5), and cyan green (superposition of blue and green): Candidatus Accumulibacter phosphatis. Figures c and f, blue: Eubacteria, purple (superposition of blue and red): Candidatus Accumulibacter phosphatis Type I, and cyan green (superposition of blue and green): Candidatus Accumulibacter phosphatis Type II.

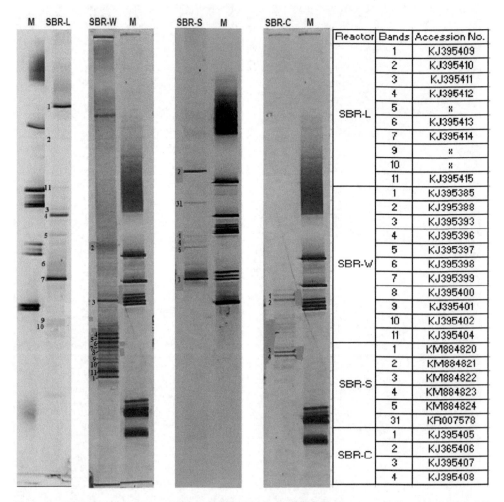

Reactor	Bands	Accession No.
SBR-L	1	KJ395409
	2	KJ395410
	3	KJ395411
	4	KJ395412
	5	x
	6	KJ395413
	7	KJ395414
	9	x
	10	x
	11	KJ395415
SBR-W	1	KJ395385
	2	KJ395388
	3	KJ395393
	4	KJ395396
	5	KJ395397
	6	KJ395398
	7	KJ395399
	8	KJ395400
	9	KJ395401
	10	KJ395402
	11	KJ395404
SBR-S	1	KM884820
	2	KM884821
	3	KM884822
	4	KM884823
	5	KM884824
	31	KR007578
SBR-C	1	KJ395405
	2	KJ365406
	3	KJ395407
	4	KJ395408

Figure S2.3 DGGE banding patterns of SBR-L, SBR-W, SBR-S and SBR-C. The sequence quality of band 5, 9 and 10 (SBR-L) was not sufficient enough to submit these sequences to the Genbank database.

3

Metabolic response of 'Candidatus Accumulibacter phosphatis' clade II to changes in the influent P/C ratio

Abstract

The objective of this study was to investigate the ability of a culture highly enriched with the polyphosphate-accumulating organism, '*Candidatus* Accumulibacter phosphatis' clade II, to adjust their metabolism to different environmental phosphate concentrations. In six different experimental phases, the biomass was cultivated in a sequencing batch reactor with acetate at different phosphate/carbon influent ratios. Activity tests were conducted to determine the anaerobic kinetic and stoichiometric parameters as well as the composition of the microbial community. Increasing influent phosphate concentrations led to increased poly-phosphate content and decreased glycogen content of the biomass. Faster active biomass specific phosphate release rates were observed as biomass poly-phosphate content increased. Together with the phosphate release rates, acetate uptake rates also increased up to an optimal poly-phosphate/glycogen ratio of 0.3 P-mol/C-mol at which the organisms were most competitive. At higher poly-phosphate/glycogen ratios, the acetate uptake rates started to decrease. The stoichiometry of the anaerobic conversions clearly demonstrated a metabolic shift from a glycogen domiated to a poly-phosphate dominated metabolism as the biomass poly-phosphate content increased. FISH and DGGE analyses confirmed that no significant changes occurred in the microbial community, suggesting that the changes in the biomass activity were due to different metabolic behavior, allowing the organisms to proliferate under conditions with fluctuating phosphate levels.

Submitted as: Welles, L., Abbas, B., Sorokin, D.Y., Lopez-Vazquez, C. M., Hooijmans, C. M., van Loosdrecht M.C.M., Brdjanovic, D. (submitted). Metabolic response of '*Candidatus* Accumulibacter phosphatis' clade II to changes in the influent P/C ratio. *Water research*

3.1 Introduction

To prevent the receiving waters from eutrophication, the Enhanced Biological Phosphorus Removal (EBPR) process is a cost-effective and environmentally-friendly process for phosphorus removal in wastewater treatment activated sludge systems. The organisms responsible for EBPR are called phosphate-accumulating organisms (PAO). Under anaerobic conditions, PAO are able to take up volatile fatty acids (VFA), such as acetate (HAc) and propionate (HPr), and store them intracellularly as poly-β-hydroxyalkanoates (PHA) (Wentzel et al., 1985, Comeau et al., 1986, Mino et al., 1987). The uptake and storage processes requires energy and reducing power. Cleavage of intracellular polyphosphate (poly-P), and subsequent release of ortho-phosphate into the liquid phase, is assumed the main pathway for all energy generation. Glycolysis of intracellular glycogen is considered the main reducing power source (Mino et al., 1987; Smolders et al., 1994). However, several studies have shown that PAO glycolysis in PAO can also function as an alternative energy generating pathway when the energy production pathway from poly-P is limiting (Brdjanovic et al., 1998; Zhou et al., 2008; Erdal et al., 2008; Hesselman et al., 2000; Acevedo et al., 2012; Welles et al., 2014). Under aerobic conditions, PAO are able to grow and to take up and store ortho-phosphate in excess as intracellular polyphosphate, leading to P-removal from the bulk liquid by wastage of activated sludge (Mino et al., 1998).

The stoichiometry and kinetic rates of EBPR anaerobic conversions are still controversial. Reported anaerobic P-release/HAc-uptake ratios range from 0.15 up to 0.93 P-mol/C-mol (Pereira et al., 1996; Hesselman et al., 2000; Smolders et al., 1994; Wentzel et al., 1987; Kisoglu et al., 2000) and kinetic rates range from 1 up to 7 (C-mmol HAc/(gVSS.h)) (Smolders et al., 1994; Liu et al., 1997; Sudiana et al., 1999; Filipe et al., 2001; Schuler and Jenkins 2003b). Such differences have often been explained based on differences on microbial composition and operational conditions. Presence of glycogen-accumulating organisms (GAO) has been suggested as a microbial factor affecting the anaerobic stoichiometry (Mino et al., 1987). GAO compete with PAO for substrate but neither release phosphate anaerobically nor store it under aerobic conditions. Besides GAO, the enrichment of different PAO clades is another factor since PAO I mainly relies on poly-P as energy source for VFA uptake, while PAO II utilizes a mixed PAO-GAO metabolism where glycogen generates a significant part of the energy required for VFA uptake (Welles et al., *submitted*). Regarding the operational conditions, pH, nature of the carbon source and calcium concentration affect the anaerobic stoichiometry (Smolders et al., 1994; Barat et al.,2006; Barat et al., 2008).

In addition, it has been suggested that the PAO biomass P-content affects the anaerobic stoichiometry. In this regard, at short-term, enriched PAO cultures can shift from a poly-P dependent towards a glycogen dependent metabolism when the biomass P-content decreases (Brdjanovic et al., 1998; Hesselman et al., 2000; Zhou et al., 2008; Erdal et al., 2008; Acevedo et al., 2012). Using highly enriched PAO I and II cultures, Welles et al. (submitted) demonstrated that the poly-P depletion led to a seven and two-fold decrease in the anaerobic kinetic rates of PAO I and II, respectively. Although these studies clearly indicated that the PAO possess metabolic flexibility, the results only represent the short-term response to changes in the biomass P-content. Furthermore, in most of the studies the specific PAO clades were either not reported or the biomass consisted of a mixture of different PAO clades (Brdjanovic et al., 1998; Hesselman et al., 2000; Zhou et al., 2008; Erdal et al., 2008; Acevedo et al., 2012). Finally, not all studies covered a broad range of different biomass P-contents (Welles et al., submitted). Therefore, it remains unclear how the kinetic rates and stoichiometry of specific PAO clades I and II are affected by a wide range of different biomass P-contents during long-term operation.

In two long-term studies, Liu et al. (1997) and Schuler and Jenkins (2003a) observed a gradual shift from a PAO metabolism to a GAO metabolism when the P-influent concentrations and intracellular poly-P content decreased. In particular, Schuler and Jenkins (2003b) observed higher HAc-uptake rates when the metabolism shifted from a GAO- to a PAO-metabolism. To explain their observations, Liu et al. (1997) suggested that a PAO-GAO competition might have taken place; whereas, Schuler and Jenkins (2003) suggested that the PAO and GAO metabolisms could be either two unique metabolisms in separate groups of organisms or two components of one metabolism in one single group of organisms. In those studies; however, no microbial identification analyses were performed. Kong et al. (2002), in a similar study, observed a shift from β-*Proteobacteria* (the subdivision to which most '*Candidatus* Accumulibacter phosphatis' belong) at a high P/C influent ratio to *a-Proteobacteria* (the subdivision to which '*Defluviicoccus*' belong) and *γ-Proteobacteria* (the subdivisions to which most '*Candidatus* Competibacter phosphatis' belong) at a low P/C influent ratio. Recent studies have demonstrated that PAO I and II have very different characteristics in terms of morphology, stoichiometry and kinetic rates, the ability to denitrify and possibly tolerance to stress conditions as well (Carvalho et al., 2007; Flowers et al., 2009; Slator et al., 2010, Welles et al., submitted). These differences may lead to prevalence of specific clades under certain conditions and consequently differences in the metabolic conversions. Considering the metabolic differences, prevalence of specific clades may significantly affect EBPR processes. For instance, differences in the anaerobic stoichiometry may affect the P-release efficiency in combined chemical and biological P-removal and P-recovery processes. Therefore it becomes important to study the metabolism of the specific clades separately to get a better understanding about the conditions favouring the specific clades. Furthermore, considering the big differences (four times bigger at poly-P depleted conditions) in kinetic rates (Welles et al., submitted), the different PAO clades can no longer be considered as one organism in modelling approaches.

Therefore, the objective of this study was to assess at long-term how the anaerobic kinetics and stoichiometry of a highly enriched PAO II culture are affected by a wide range of different biomass P-contents. From a fundamental perspective, this will provide a more quantitative insight into the relationship between storage polymers and anaerobic metabolic pathways, contributing to explain the wide range of different P/HAc ratios and kinetic rates observed in previous studies. From a practical perspective, this important understanding will ultimately help to improve the existing metabolic models, leading to better design and operation of the EBPR processes and, in particular, of combined chemical and biological phosphorus removal and recovery systems.

3.2 Material and Methods

3.2.1 Experimental phases
Using a lab scale sequencing batch reactor (SBR), a PAO culture was enriched as described in previous studies (Welles et al., 2014; Welles et al., accepted). To investigate the effect of the biomass P-content on the kinetic rates and stoichiometry of the anaerobic conversions, six long-term experimental phases were designed with the aim to obtain an EBPR biomass with different P-contents. In phase 0, the phosphate concentration was 20mgP/L. The influent phosphate concentrations were decreased in phase 1, increased in phase 2 and 3, decreased in phase 4 and finally increased in phase 5, to see if possible changes in the biomass activity were reversible. With regard to any other phase, the only difference in the operational conditions was the influent ortho-phosphate concentration supplied: 20, 15, 30, 45, 15 and 20 mgP/L in phases 0, 1, 2, 3, 4 and 5, respectively. The length of each phase was 782, 143, 22, 18, 29 and 49 days for phases 0, 1, 2, 3, 4 and 5, respectively.

3.2.2 SBR operation
The PAO culture was enriched in a 2.5 L double-jacketed laboratory SBR, operated and controlled automatically in a sequential mode by an Applikon ADI controller also used for data acquisition and

storage (e.g. pH and O_2) using BioXpert software (Applikon, The Netherlands, Schiedam). The reactor was inoculated with activated sludge from a municipal wastewater treatment plant with a 5-stage Bardenpho configuration (Hoek van Holland, The Netherlands).

The SBR was operated in cycles of 6 hours (2.25 h anaerobic, 2.25 aerobic and 1.5 settling phase) following similar operating conditions used in previous studies (Welles et al., 2014; Smolders et al., 1994; Brdjanovic et al., 1997). The pH was maintained at 7.0 by dosing 0.4 M HCl and 0.4 M NaOH and temperature was controlled at 20 ± 1 °C. Each cycle started with a 5 min sparging phase with nitrogen gas at a flow rate of 30 L/h to create anaerobic conditions. After the first 5 min, 1.25 L of synthetic substrate was fed to the SBR over a period of 5 min and nitrogen gas sparging continued throughout the anaerobic phase. In the aerobic phase, compressed air was sparged to the SBR at a flow rate of 60 L/h. Mixing was provided at 500 rpm, except during settling and decant phases when mixing was switched off.

The SBR was controlled at a biomass retention time (SRT) of 8 days, not taking into account the potential loss of solids in the effluent during removal and biofilm removal during regular cleaning. At the end of the settling period, supernatant was pumped out from the reactor, leaving 1.25 L of mixed liquor in the reactor. This resulted in a total hydraulic retention time (HRT) of 12 h.

3.2.3 Medium
The SBR was fed with the same medium in all the experimental phases. The only difference was the orthophosphate concentration. The respective orthophosphate concentrations (provided with the addition of $NaH_2PO_4.H_2O$) were: (phase 0) 0.64 P-mmol/L (influent P/C ratio is 0.051 P-mol/C-mol) ; (phase 1) 0.48 P-mmol/L(influent P/C ratio is 0.038 P-mol/C-mol) ; (phase 2) 0.96 P-mmol/L(influent P/C ratio is 0.076 P-mol/C-mol); (phase 3i) 1.44 P-mmol/L(influent P/C ratio is 0.114 P-mol/C-mol); (phase 4) 0.48 P-mmol/L(influent P/C ratio is 0.038 P-mol/C-mol); and, (phase 5) 0.64 P-mmol/L(influent P/C ratio is 0.051 P-mol/C-mol). The concentrated medium was prepared with demineralised water. In the beginning of every cycle, 250 mL of concentrated substrate together with 1000 mL demi water were fed to the reactor. After dilution, the influent contained per litre: 860 $mgNaAc.3H_2O$ (12.6 C-mmol/L, 405 mgCOD/L), 107 $mgNH_4Cl$ (2 N-mmol/L), 120 mg $MgSO_4.7H_2O$, 140 mg $CaCl_2.2H_2O$, 480 mg KCl, 2 mg of allyl-N-thiourea (ATU) to inhibit nitrification, 0.3mL/L trace element solution, and a defined concentration of $NaH_2PO_4.H_2O$ different in each experimental phase as previously described. The trace element solution was prepared as described by Smolders et al., (1994). Prior to use, both concentrated solutions were autoclaved at 110 oC during 1 h.

3.2.4 SBR monitoring
The performance of the SBR was regularly monitored by measuring ortho-phosphate (PO_4^{3-}-P), acetate (HAc), mixed liquor suspended solids (MLSS) and mixed liquor volatile suspended solids (MLVSS). Stable performance in the reactor was confirmed by daily observation of the aforementioned parameters as well as pH and DO online data. At the end of each experimental phase, (except for phase 0), a cycle test was conducted to determine the anaerobic stoichiometric and kinetic parameters. In the cycle tests, polyhydroxyalkanoate (PHA) and glycogen concentrations were determined in addition to the above described parameters. Furthermore, the composition of the microbial community was characterized by fluorescence in situ hybridization (FISH) analysis and denaturing gradient gel electrophoresis (DGGE) analysis.

3.2.5 Kinetic rates and stoichiometric values
The PO_4-release rates and HAc-uptake rates were determined using the PO_4 and HAc profiles observed in the cycle tests and expressed as maximum active biomass specific rates as described by Smolders et al.

(1994) and Brdjanovic et al. (1997). The stoichiometric parameters of interest were: P/HAc, PHV/HAc, PHB/HAc, PHV/PHB, gly/HAc and gly/PHB.

3.2.6 Analyses

Determination of TSS, VSS and PO_4^{3-}-P concentrations were performed in accordance with Standard Methods (A.P.H.A., 1995). HAc was determined using a Varian 430-GC Gas Chromatograph (GC) equipped with a split injector (split ratio 1:10), a WCOT Fused Silica column with a FFAP-CB coating (25 m x 0.53mm x 1μm), and coupled to a FID detector. Helium gas was used as carrier gas. Temperature of the injector, column and detector were 200°C, 105 °C and 300°C, respectively. PHB and PHV content of freeze dried biomass was determined according to the method described by Smolders *et al.* (1994). Glycogen was determined according to the method described by Smolders *et al.* (1994) but with an extended digestion of 5 h in 5 mL 0.9M HCl, using 5 mg of freeze-dried biomass as described by Lanham et al. (2012).

3.2.7 Characterization of microbial populations

An estimation of the degree of enrichment of the bacterial populations of interest (PAO Type I, PAO type II and GAO) was based on FISH analyses, following the procedure described by Winkler et al. (2011). All bacteria were targeted by the EUB338mix (general bacteria probe) (Amann *et al.*, 1990; Daims *et al.*, 1999). 'Candidatus Accumulibacter phosphatis' and '*Candidatus* Competibacter phosphatis' were targeted by PAOMIX probe (mixture of probes PAO462, PAO651 and PAO846) (Crocetti *et al.*, 2000) and GAOMIX probe (mixture of probes GAOQ431 and GAOQ989) (Crocetti *et al.*, 2002), respectively. PAO I (clade IA and other type I clades) and PAO II (clade IIA, IIC and IID) were targeted by the probes Acc-1-444 and Acc-2-444 (Flowers *et al.*, 2009), respectively. The quantification of the PAO fractions in the biomass was carried, following the procedures described by Welles et al. (2014) using around 16 randomly selected separate images for each quantification.

To confirm the FISH observations and to identify potential changes in the microbial populations at the sub-clade level, 16S-rDNA-PCR DGGE was applied. Samples were collected at the end of each experimental phase. DNA extraction, PCR amplification, DGGE, band isolation, sequencing and identification of microorganisms were carried out according to the procedures described by Bassin *et al.* (2011). Two double confirm the specific PAO clade, ppk1 gene fragments were recovered and analysed, following the procedures described by McMahon et al. (2007), using the pimers ACCppk1-254F en ACCppk1-1376R for PCR amplification.

3.2.8 Microscopy

For thin sectioning electron microscopy, the cells were first fixed in 3% (v/v) glutaraldehyde for 1 h on ice, then postfixed in 1% (w/v) OSO_4 + 0.5 M NaCl for 3 h at room temperature, washed and stained overnight with 1% (w/v) uranyl acetate, dehydrated in ethanol series and embedded in Epoxy resin. The thin sections were finally stained with 1% lead acetate.

3.2.9 Active biomass

The active biomass concentration was determined as described in the section 2.2.4 of the previous chapter.

3.2.10 Estimation of Poly-P

The concentration of PAO Poly-P was estimated on the basis of the ISS/TSS ratio and confirmed using steady-state mass balances as described in Welles et al. (2015). Equation 1 was developed using the ISS/TSS ratio of the biomass, assuming that (i) the ISS/TSS ratio associated with active biomass in non-EBPR biomass (ISS_b) was 0.025 mg ISS/mgTSS (as observed in this study after poly-P depletion), (ii) a poly-P composition of $(PO_3)_3MgK$ with a P-content ($f_{P,ppASH}$) of 0.31mgP/mgISS and, (iii) negligible chemical precipitation. Equation 2, derived from the steady-state mass balance of phosphorus, assuming that (i) the solids in the effluent were negligible, (ii) a ratio of non poly-P phosphorus per VSS ($f_{P,bVSS}$) equal to the P-content of non-EBPR biomass at around 0.023 mgP/mgVSS (Metcalf and Eddy, 2003) and, (iii) absence of chemical precipitation. A detailed description of the development of the equations can be found in the Appendix.

poly-P = (ISS - $f_{ISSb,TSS}$, / (1- $f_{ISSb,TSS}$) * VSS) * $f_{P,ppISS}$ (Eq. 1)

poly-P = SRT/HRT * ($T_{P,i}$ - $T_{P,e}$) - $f_{P,bVSS}$ * VSS   (Eq. 2)

where;

VSS: Volatile suspended solids

ISS: Inorganic suspended solids

Poly-P: Poly-phosphate

$T_{P,i}$: Total phosphorus concentration in the influent

$T_{P,e}$: Total phosphorus concentration in the effluent

$f_{P,bVSS}$: Ratio of non poly-P phosphorus per VSS

$f_{ISSb,TSS}$: ISS/TSS ratio associated with active biomass

$f_{P,ppISS}$: P-content of poly-P

HRT: Hydraulic retention time

SRT: Solids retention time

3.3 Results

3.3.1 EBPR performance at different P/C influent ratios

An enriched EBPR culture was cultivated with different influent ortho-phosphate concentrations during six experimental phases (phases 0 to 5). The increase in influent phosphate concentration resulted in higher P-release and consequently higher ortho-phosphate concentrations at the end of the anaerobic phase (Figure 3.1a) and higher biomass ISS/TSS ratio at the end of the aerobic phase (Figure 3.1b). A decrease in the influent phosphate concentrations decreased both the biomass ISS/TSS ratio content and ortho-phosphate concentrations at the end of the anaerobic phase, indicating that the observed patterns were reversible. The P-release/HAc-uptake ratio fluctuated between 0.15 and 0.6 P-mol/C-mol, depending on the influent ortho-phosphate concentration, showing the highest P-release in phase 4 at the highest influent ortho-phosphate concentration. Full P-removal was observed in all phases except in phase 4 in which an average ortho-phosphate concentration of 0.20 P-mmol/L was measured in the effluent.

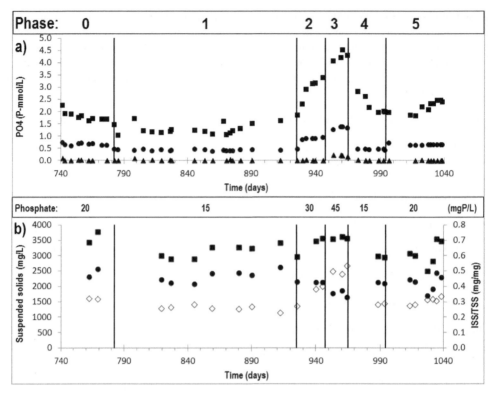

Figure 3.1 Concentrations observed in the different cycles of the experimental phases studied: (a) ortho-phosphate concentrations in the influent (●), end of the anaerobic phase (■), and end of the aerobic phase (▲); and, (b) TSS concentrations in the end of the aerobic phase (■), VSS concentrations in the end of the aerobic phase (●) and ISS/TSS ratios (◇).

3.3.2 Microbial community at different influent P/C ratios

FISH analysis and their quantification showed that PAO were highly dominant in experimental phase 1 (comprising up to 99±3% PAO) and more specifically that the PAO population consisted of PAO clade II with minor traces of PAO clade I (99±6% PAO II and 1±2% PAO I)(Figure 3.2a and Figure 3.2b). GAO were not detected and only minor traces of PAO clade I were observed. Based on FISH analyses, no significant changes occurred in the relative quantities of PAO I, PAO II and GAO throughout the execution of the experimental phases (Figure 3.2b, c, e and f) and therefore quantification was not considered. 16S-rRNA gene based DGGE profiles were obtained from samples collected at the end of each experimental phase (Figure 3.2g). From the DNA derived DGGE patterns, 24 bands were selected, covering all the unique bands observed in the different phases. The phylogenetic analysis of selected bandsequences was conducted (Table 3.1. The bacterial groups detected were: (i) β-proteobacteria closely related to 'Candidatus Accumulibacter phosphatis' Type IIC/D (bands 7, 21 and 22); (ii) β-proteobacteria not closely related to 'Candidatus Accumulibacter phosphatis' (band 9); (iii) δ-proteobacteria (band 13); (iv) α-proteobacteria (bands 06 and 04); (v) Armatimonadetes were also detected (band 20), as well as (vi) bacteria belonging to Bacteroidetes (bands 14, 17, 2, 1, 15, 18, 16, 12, 3 and 11). Bands 05, 08, 10, 19, 23, and 24 could not provide sufficient DNA of the required quality for sequencing and the quality of the sequence obtained from band 9 was insufficient for submission to the genebank. . In a in a previous study (Welles et al., 2015), a phylogenetic tree was constructed on the basis of the 16S rRNA-gene sequences obtained from the biomass sample collected at the end of experimental phase 1. An additional ppk1 gene fragment analysis conducted in this study revealed that gene fragments obtained from the biomass in experimental stage 5, were closely related (99.7%) to a PAO strain obtained in a previous study, BA91 (Acc.no. JDVG02000730) (Skennerton at al., 2014), that was classified as 'Candidatus Accumulibacter phosphatis' clade IIC, confirming the observations from the 16S rRNA-gene based DGGE analysis in this study.

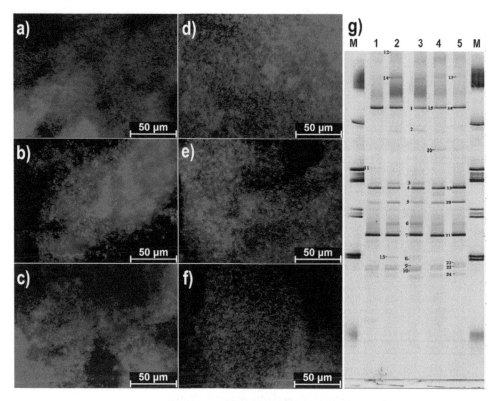

Figure 3.2 Representative FISH microscopic images showing the distribution of bacterial populations in biomass samples collected at the end of phase 1 (a and d), phase 3 (b and e) and phase 5 (c and f) on day 890, 964 and 1034, respectively. In Figures a, b and c; blue: EUB mix (Cy5); purple (superposition of blue and red): PAO mix (Cy3); and cyan green (superposition of blue and green): GAO mix (Fluos). In Figures d, e and f; blue: PAO mix (Cy5), purple (superposition of blue and red): PAO clade II (Cy3), and cyan green (superposition of blue and green): PAO clade I (Fluos). 16S rRNA bands obtained by DGGE (g). M = marker and Number 1 to 5 correspond to biomass samples collected at the end of the experimental phases on day 890 (phase 1), day 945 (phase 2), day 964 (phase 3), day 993 (phase 4), and day 1034 (phase 5). Phylogenetic analysis of selected bands is shown in Table 3.1.

Among all sequences, neither 'Candidatus Competibacter phosphatis' nor Defluviicoccus (another known GAO) were detected. Neither were sequences detected that were related to Gammaproteobacteria, the subdivision to which most GAO belong. In the different experimental phases, no significant differences were observed in the dominant bands (band 1, 4 and 7), representing bacteria closely related to Bacteroidetes, 'Candidatus Accumulibacter phosphatis' type IIC/D and a-proteobacteria, respectively. From the minor bands, band 3, representing Bacteroidetes (closely related to Flavihumibacter petaseus) and band 6, representing an a-proteobacteria (closely related to Rhodobacter capsulatus), showed slightly higher intensity during phases 2 and 3 which were the phases with higher influent ortho-phosphate concentrations (30 and 45 mgP/L, respectively), but these bands were not detected when the influent ortho-phosphate concentration decreased in phase 4 and 5 to 15 and 20 mgP/L , respectively.

RDP classification					
Band	Accession No.	Phylum	Class	Order	Family / (Suborder)	genus
1	KJ395409	Bacteroidetes	Sphingobacteriia	Sphingobacteriales	Chitinophagaceae	-
12	KJ395416	Bacteroidetes	Sphingobacteriia	Sphingobacteriales	Chitinophagaceae	-
15	KJ395419	Bacteroidetes	Sphingobacteriia	Sphingobacteriales	Chitinophagaceae	-
18	KJ395422	Bacteroidetes	Sphingobacteriia	Sphingobacteriales	Chitinophagaceae	-
3	KJ395411	Bacteroidetes	Sphingobacteriia	Sphingobacteriales	Chitinophagaceae	Flavihumibacter
11	KJ395415	Bacteroidetes	Sphingobacteriia	Sphingobacteriales	Chitinophagaceae	unclassified_Chitinophagaceae
16	KJ395420	Bacteroidetes	Sphingobacteriia	Sphingobacteriales	Chitinophagaceae	Terrimonas
2	KJ395410	Bacteroidetes	Flavobacteriia	Flavobacteriales	Flavobacteriaceae	Chryseobacterium
14	KJ395418	Bacteroidetes	Flavobacteriia	Flavobacteriales	Flavobacteriaceae	Flavobacterium
13	KJ395417	Proteobacteria	Deltaproteobacteria	Myxococcales	(Cystobacterineae)	-
21	KJ395424	Proteobacteria	Betaproteobacteria	Rhodocyclales	Rhodocyclaceae	Propionivibrio
22	KJ395425	Proteobacteria	Betaproteobacteria	Rhodocyclales	Rhodocyclaceae	Propionivibrio
7	KJ395414	Proteobacteria	Betaproteobacteria	Rhodocyclales	Rhodocyclaceae	Propionivibrio
4	KJ395412	Proteobacteria	Alphaproteobacteria	-	-	-
6	KJ395413	Proteobacteria	Alphaproteobacteria	Rhodobacterales	Rhodobacteraceae	-
20	KJ395423	Armatimonadetes	-	-	-	Armatimonadetes_gp5

BLAST results				
Band	Accession No.	Closest relative (excl. uncultured/env. samples)	ACC	% similarity
1	KJ395409	Trachelomonas volvocinopsis var. spiralis strain UTEX1313	FJ719709	98
12	KJ395416	Trachelomonas volvocinopsis var. spiralis strain UTEX1313	FJ719709	96
15	KJ395419	Trachelomonas volvocinopsis var. spiralis strain UTEX1313	FJ719709	98
18	KJ395422	Trachelomonas volvocinopsis var. spiralis strain UTEX1313	FJ719709	98
3	KJ395411	Flavihumibacter solisilvae strain 3	KC569790	97
11	KJ395415	Heliimonas saccharivorans strain L2-4	JX458466	87
16	KJ395420	Terrimonas lutea DY	NR_041250	97
2	KJ395410	Chryseobacterium sp. Y1D	EU839047	98
14	KJ395418	Flavobacterium croceum strain EMB47	NR_043768	99
13	KJ395417	Cystobacter sp. GNDU S198	KP178619	95
21	KJ395424	Candidatus Accumulibacter phosphatis clade IIA str. UW-1	NR_074763	98
22	KJ395425	Candidatus Accumulibacter phosphatis clade IIA str. UW-1	NR_074763	95
7	KJ395414	Candidatus Accumulibacter phosphatis clade IIA str. UW-1	NR_074763	98
4	KJ395412	Rhizobium giardinii CCNWSX1555	KP875539	92
6	KJ395413	Rhodobacter sp. EMB 174	DQ413163	98
20	KJ395423	Fimbriimonas ginsengisolii Gsoil348	CP007139	86

Table 3.1 Phylogenetic analysis of 16S rRNA sequences obtained from DGGE band profiles shown in Figure 3.2.

3.3.3 Effect of influent P/C ratio on intracellular storage polymers

The effect of the influent P/C ratio on the ISS/TSS ratio of the biomass and storage polymers is shown in Figure 3.3a and 3.3b. As the influent P/C ratio increased, the ISS/TSS ratio increased in a linearly proportional fashion. The data point belonging to the influent P/C ratio that corresponds to zero was obtained in a batch experiment where the poly-P was depleted from the enriched biomass. At higher biomass ISS/TSS ratios, the poly-P/active biomass ratio and poly-P/gly ratio increased while the glycogen/active biomass ratio decreased. Apparently, PAO favor the storage of poly-P over glycogen at high P influent. To get a better understanding of the intracellular organization of storage polymers, thin sections were prepared from the PAO cells taken at the end of the anaerobic phase in experimental phase 2 and analysed by electron microscopy (Figure 3.3c). In this phase, with an influent phosphorus concentration of 30 mg/L the biomass ISS/TSS ratio reached about 0.4 at the end of the aerobic phase. The microscopy showed that poly-P appeared as 1 or 2 large dense inclusions. Smaller electron transparent inclusions surrounded by a membrane were observed and assumed to correspond to PHA, as it is known to be surrounded by a membrane (Liebergesell et al., 1994; Pieper-Furst et al., 1994; Steinbuchel et al., 1995; Mayer et al., 1996) while the white dispersed spots were considered to be glycogen, which is known to be freely dispersed in the cytosol (Brana et al., 1980; Kamio et al., 1981).

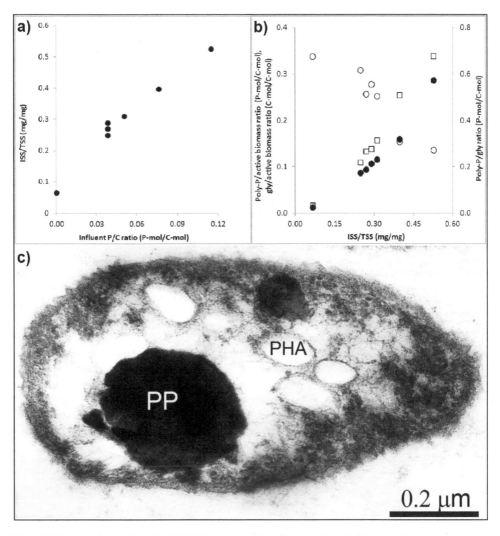

Figure 3.3 Storage polymers in enriched PAO biomass samples collected at the end of the experimental phases on day 881 and 890 (phase 1), day 945 (phase 2), day 964 (phase 3), day 993 (phase 4), and day 1034 (phase 5): (a) ash/TSS ratio as a function of the influent P/C ratio at the end of the aerobic phase; (b) the relationship between the ash/TSS ratio at the end of the aerobic phase and (●) estimated poly-P/active biomass ratio, (○) glycogen/active biomass ratio and (□) poly-P/gly ratio; and, (c) electron microscope image of a thin section showing the poly-P, PHA and glycogen organization in the enriched PAO II cell at the end of the anaerobic phase.

3.3.4 Effect of P-content on PAO kinetics

During acetate uptake, the specific P-release rates increased strongly when the poly-P/active biomass ratio increased from 0 to 0.2 P-mol/C-mol, above which the rate seemed to level off (Figure 3.4a). Also, the endogenous P-release rate increased with the increasing poly-P/active biomass ratio. When the P-release rate increased, the HAc-uptake rate also increased. A maximum HAc-uptake rate of 0.20 C-mol/C-mol.h was observed when the P-release rate reached 0.07 P-mol/C-mol.h at a poly-P/gly ratio of around 0.3 P-mol/C-mol. Above this poly-P/gly ratio, the HAc-uptake rate decreased.

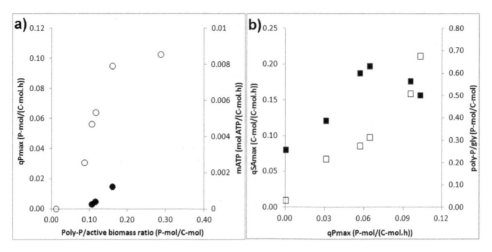

Figure 3.4 Effects of poly-P contents and P-release rates on specific biomass kinetic rates at the end of the experimental phases on day 881 and 890 (phase 1), day 945 (phase 2), day 964 (phase 3), day 993 (phase 4), and day 1034 (phase 5): (a) specific P-release rates during HAc uptake (○: qPmax) and endogenous maintenance activity (●: mATP) as a function of the poly-P/active biomass ratio; and, (b) HAc-uptake rates (■: qSAmax) and poly-P/gly ratio (□: poly-P/gly) versus P-release rates.

3.3.5 Effect of P-content on PAO stoichiometry

A clear relationship was observed between the P-content and stoichiometry of the anaerobic conversions. The PO_4-release/HAc-uptake ratio increased drastically when the poly-P/active biomass ratio increased (Figure 3.5a), suggesting that the energy production by poly-P consumption per HAc-uptake increased at higher poly-P/active biomass ratios. Consequently, the gly/HAc ratio, PHV/gly ratio, PHV/HAc ratio, PHB/HAc ratio and PHV/PHB ratio decreased when the PO_4/HAc ratio increased (Figure 3.5b-f), which indicates a shift in the metabolic pathways of PAO for the generation of energy from a glycogen dependent metabolism to a poly-P dependent metabolism.

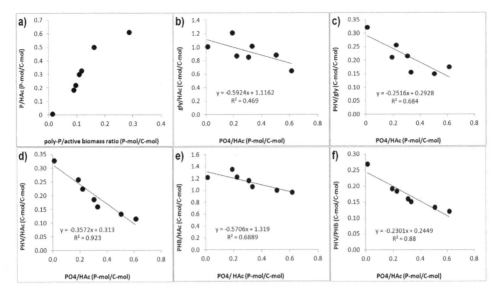

Figure 3.5 Effect of poly-P content on anaerobic PAO stoichiometric parameters at the end of the experimental phases on day 881 and 890 (phase 1), day 945 (phase 2), day 964 (phase 3), day 993 (phase 4), and day 1034 (phase 5): (a) PO_4/HAc ratio versus estimated poly-P/active biomass ratio; (b) gly/HAc ratio versus PO_4/HAc ratio; (c) PHV/gly versus PO_4/HAc ratio; (d) PHV/HAc versus PO_4/HAc ratio; (e) PHB/HAc versus PO_4/HAc ratio; and, (f) PHV/PHB ratio versus PO_4/HAc ratio.

3.4 Discussion

3.4.1 Effect of influent P/C ratio on EBPR performance, storage polymers and microbial population dynamics.

At the lowest influent P/C ratio in experimental phase 1, the stoichiometry of the anaerobic conversions, i.e. P/HAc ratio of 0.2 P-mol/C-mol, indicated that the biomass exhibited a mixed PAO and GAO metabolism. However, the population was highly dominated throughout the different experimental phases by 'Candidatus Accumulibacter phosphatis' clade II. The genus Defluviicoccus, 'Candidatus Competibacter phosphatis' (both belonging to the GAO group) and 'Candidatus Accumulibacter phosphatis' clade I, were not observed by DGGE analysis.'CanThe latter two were neither observed by FISH analysis. This indicates that at this influent P/C ratio, the PAO II performed a mixed PAO-GAO metabolism, which is in line with previous observations (Welles et al., 2014). The influent P/C ratio clearly affected the performance of the biomass in the anaerobic phase, the ISS/TSS ratio and the storage polymers. Since no significant changes occurred in the composition of the microbial community during the experimental phases, it can be concluded that the observed changes in the biomass behavior are due to a metabolic shift of 'Candidatus Accumulibacter phosphatis' clade II. Furthermore, as the poly-P content of the biomass increased, the glycogen content decreased, indicating that poly-P is the preferred storage polymer by PAO II which results in higher poly-P/glycogen ratios when the influent phosphate concentrations becomes less limiting at higher influent P/HAc ratios. Still glycogen formation was never completely eliminated.

It is remarkable that the DGGE profiles showed two more intense band patterns belonging to *Bacteriodetes* and *a-Proteobacteria* in addition to the 'Candidatus Accumulibacter phosphatis' clade IIC/D band pattern, while FISH analysis demonstrated that all flocs were highly dominated by PAO (99%). This may be explained by the differences in the analytical methods. The intensity of the DGGE bands reflects the abundance of PCR amplified 16s rRNA gene copy numbers, while bacterial fractions determined by FISH quantification represent the relative surface areas of bacterial flocs in which the target 16S rRNA is present. A more detailed microscopic analysis of the PAO II dominated flocs (data not shown) showed that in each PAO floc there were other bacteria of a smaller size homogeneously enmeshed. The smaller bacteria were less abundant on a volume or surface area basis (hardly observed by FISH), but on a cell-counting basis significantly present. Therefore, the bacterial floc's surface areas that stained positive with PAO mix probes and the EUB mix probes were very similar, leading to estimated PAO fraction close to 100%. Consequently other bacteria such as *Bacteroidetes* and *a-Proteobacteria*, were not detected by FISH analysis as a significant side populations. In the DGGE analysis the DNA is extracted from the biomass. Considering the large difference in size of the bacteria (about one order of magnitude) and the fact that each bacteria has a genome, the fraction of extracted template DNA belonging to *Bacteroidetes* and *a-Proteobacteria* in comparison to the fraction of extracted template DNA belonging to PAO may have been much larger than the dry weight biomass fractions. Other factors that may have contributed to the differences in the bacterial quantities obtained from FISH and DGGE data between these organisms may have been: (i) differences in DNA extraction efficiency, (ii) different copy numbers of the 16S rRNA gene in the genomes and, (iii) differences in PCR amplification efficiency of the 16S rRNA genes. In anyway, the discussed two groups are most probably minor and are not important for the major function of the biomass.

3.4.2 Effect of storage polymers on kinetic rates and stoichiometry

The shift in the stoichiometry of the anaerobic conversions for HAc-uptake indicated that the changes in kinetic P-release and HAc-uptake rates were associated with a shift in the relative ratio of the metabolic fluxes from the different energy generating pathways (glycogen conversion and poly-P degradation), which may be triggered by a change in the type of storage polymers. In addition, the change in the the anaerobic P-release rate for maintenance energy production (determined in the absence of acetate) seemed to increase when the biomass P-content increased. Considering the the maintenance energy requirements of microorganisms are more or less constant under defined operational conditions, the change in the P-release rate for maintenance energy production suggest that here also a metabolic shift occurred. Possibly, the changes in the P-release activity for anaerobic maintenance energy production may have been compensated by changes in glycogen consumption, but no glycogen measurements were conducted to verify this hypothesis for the anaerobic maintenance energy production as these measurements are not very accurate and the potential glycogen consumption for maintenance energy production is very little. The observed increase in the P-release rate for HAc-uptake and anaerobic maintenance energy production, in response to increased poly-P content suggests that poly-P is the preferred or most easily accessible source of energy during anaerobic HAc uptake, provided that the poly-P content of the biomass is high. When the poly-P content of the biomass decreases, the rate of energy production from poly-P consumption becomes limited, which then needs to be topped up by energy production from glycogen conversion. The consumption of Poly-P allows the cell to harvest 1 mol ATP/P-mol Poly-P (Groenestijn et al., 1987, Smolder et al., 1994), while glycogen conversion generates 0.5 mol ATP/C-mol glycogen (Zeng et al., 2003).

During the experimental phase 1 performed with the lowest influent P/C ratio (influent concentration 15mgP/L), the estimated available intracellularly stored poly-P was 5.1 P-mmol/L. This concentration was about four times higher than the poly-P actually used for HAc uptake (1.3 P-mmol/L), which indicates that the gradual shift in the metabolism of PAO is not driven by a stoichiometric limitation of the available poly-P. Considering that poly-P is stored as large inclusions (Figure 3.3c), the rate of poly-P consumption may indeed be surface area related. Similar to PHA, the surface area plays an important role in the conversion rates (Smolders et al., 1995; Murnleitner et al., 1997). If the rate depends on the surface area of the poly-P granules, then a decreased poly-P content would lead to a lower ATP formation rate.

A schematic presentation of the mechanism regulating the metabolic shift is shown in Figure 3.6. When the P-release rates increased, the HAc-uptake rates seemed to increase up to a maximum value above which the HAc-uptake rate started to decrease again. The increase in the HAc-uptake rate coupled to an increase in the P-release rate suggests that the energy production could be the rate limiting step for HAc-uptake and that at high poly-P content the energy production from poly-P is faster than the energy production from glycogen. Consequently, the HAc-uptake rate increases when both the ortho-phosphate release rate and its associated energy production rate increase. Differences in the energy production rates from poly-P and glycogen may be explained by differences in the number of metabolic conversions of each energy generating pathway. The metabolic pathway for ATP production using poly-P requires only two biochemical conversions and subsequent release of ortho-phosphate over the membrane accompanied by counterions (Groenestijn et al., 1987). Although the exact pathway for energy production from glycogen remains unclear, such as the type of glycolytic pathways, it probably requires at least 10 biochemical conversions (Satoh et al., 1994) to convert glycogen into PHA and avoid the net-production of reduction equivalents.

When the biomass P-content becomes high, the P-release and its associated energy production rate from poly-P conversion become high as well. However, at high biomass P-content, the glycogen content appears to be low, although glycogen consumption for anaerobic substrate uptake was never eliminated in the anaerobic conversions and therefore seems to be essential for the production of reduction equivalents, which was also observed in the study of Schuler and Jenkins (2003a). Possibly the production rate of reduction equivalents is also surface area related. Therefore, a decrease in the glycogen content may trigger a transition where the rate limiting step of the HAc-uptake process changes from the energy production rate to the production rate of reduction equivalents. Thus, at a high biomass P-content, the HAc uptake rate starts to decrease possibly due to limited supply rate of reduction equivalents by glycogen. An optimal HAc uptake rate seems to occur at a poly-P/gly ratio of 0.3 P-mol/C-mol, which corresponds with an ISS/TSS ratio of 0.3 mg ash/mg TSS obtained at an influent P/C ratio of 0.05 P-mol/C-mol. This is roughly the influent P/C ratio at which the PAO culture was originally enriched in this study.

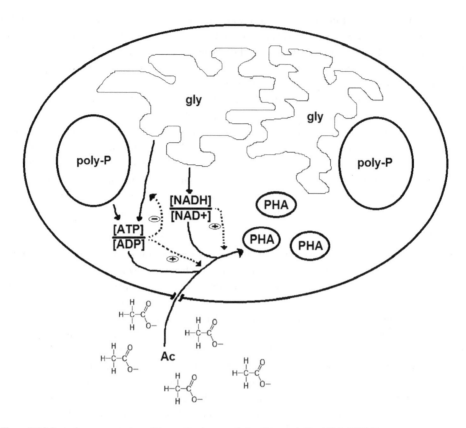

Figure 3.6 Schematic representation of the mechanism regulating the metabolic shift in PAO II.

3.4.3 Assessment of the anaerobic stoichiometric parameters against previous values reported in literature

Table 3.2 shows a comparison between the anaerobic kinetic rates and stoichiometric parameters obtained in this study and previously reported values (Liu et al., 1997; Schuler and Jenkins, 2003a,b; among others with highly enriched PAO I, PAO II and GAO cultures). Similar to the findings of Liu et al. (1997) and Schuler and Jenkins (2003b), the P/HAc stoichiometry increased when the biomass P-content increased but the dependency of the stoichiometry was different. In this study, the P/HAc stoichiometry at a medium initial estimated P/TSS ratio of 0.07mg/mg was 0.17 P-mol/C-mol against 0.32 P-mol/C-mol reported by Schuler and Jenkins (2003a). Recently, at a defined poly-P/VSS ratio (0.125 mg/mg), the anaerobic P/HAc stoichiometry of PAO I was reported to be 2.4 times higher than the stoichiometry of PAO II (Welles et al., 2015). Therefore, these findings suggest that the biomass enriched by Schuler and Jenkins. (2003a,b,c) may have been dominated by PAO I. In both studies, the maximum non-soluble phosphorus (Pns)/TSS ratio (Pns = total unfiltered phosphorus minus filtered ortho-phosphate) of the enriched cultures seem to be similar, but the stoichiometric ratio obtained at highest Pns/TSS ratio (poly-P saturated biomass obtained in reactors operated under P non-limiting conditions) was different. In the present report, the anaerobic P/HAc stoichiometry of PAO clade II increased from 0.03 to 0.65 P-mol/C-mol when the initial biomass P-content increased from 0.01 to 0.17mgP/mgTSS, while the P/HAc stoichiometry of the biomass observed by Schuler and Jenkins (2003a) increased from 0.03 to 0.72 P-mol/C-mol when the initial biomass P-content increased from 0.02 to 0.16 mgP/mgTSS.

Considering the kinetic rates, the profile of the kinetic HAc-uptake rates versus the P-release rates obtained in this study seems to be different to the observations of Liu et al. (1997) and Schuler and Jenkins (2003b). In those studies, the HAc-uptake seemed to increase proportionally when the P-release rates increased. A minimal HAc-uptake rate (0.08 C-mmol/(gVSS.h)) was observed when the P-release rate was close to zero and a maximal HAc-uptake rate (7 C-mmol/(gVSS.h)) was observed when the P-release rate reached a maximum value of 5.5 (P-mmol/(gVSS.h)). On the opposite, in this research, a considerably higher HAc-uptake rate (2.2 C-mmol/(gVSS.h)) was obtained when no P-release occurred. Moreover, a maximum HAc-uptake rate (6.0 C-mmol/(gVSS.h)) was obtained at a medium P-release rate (1.9 P-mmol/(gVSS.h)) when the biomass contained 0.095 (mgP/mgTSS). At this P-content, Schuler and Jenkins (2003b) observed a lower HAc-uptake rate (±4.0 C-mmol/(gVSS.h)). Interestingly, in this study at high P-content with the maximum P-release rate (3.2P-mmol/(gVSS.h)), a lower HAc-uptake rate (4.9 C-mmol/(gVSS.h)) was observed .

These differences in the kinetics support the hypothesis that in the study of Liu et al. (1997) the biomass was dominated by PAO I. In short-term batch experiments, Welles et al. (submitted), demonstrated that the specific HAc-uptake rate of PAO I is four times lower than the HAc-uptake rate of PAO II in the absence of poly-P. Furthermore, the HAc-uptake rate of PAO I obtained under steady-state conditions when the biomass P-content was high decreased up to seven times when the poly-P content depleted.

The findings drawn from this study, and those from Liu et al., (1997) and Schuler and Jenkins (2003), help to explain the observations of Barat *et al.* (2008) where the anaerobic P/HAc ratio decreased because, likely, the higher calcium concentrations induced the chemical precipitation of phosphorus (during the aerobic phase) reducing the available poly-P content of the biomass and leading to a lower P/HAc ratio in the anaerobic phase of their experiments.

3.4.4 PAO-GAO Competition

Assuming that the biomass in the study of Schuler and Jenkins (2003) was dominated by PAO I, the findings from this study suggest that in the low P-content range, PAO II may have a competitive advantage over PAO I and that at a high P-content PAO I may have a competitive advantage over PAO II. This was also demonstrated in short-term experiments (Welles et al., 2015). However, the kinetic rates, and their associated competitiveness of certain PAO clades (presumably clade I), increase as the P-content of the biomass increases. If, in a certain system, GAO become dominant, the availability of P per PAO biomass, and therewith the PAO specific P-content, increases. Consequently, the HAc uptake rate of PAO I and its ability to compete with GAO may increase, while that of PAO II may decrease, when the poly-P/gly ratio increases above the optimal ratio. Further research on enriched PAO I cultures is needed to validate these hypotheses.

In this study, the maximum specific HAc-uptake rate of PAO II (at 20ºC) at a medium P-content was similar to the HAc uptake rates observed with highly enriched GAO cultures (Lopez-Vazquez et al., 2007; Zeng et al., 2003). However, the VSS specific biomass HAc-uptake rate of PAO (presumably clade I) in the study of Schuler and Jenkins (2003) at the highest P-content seemed to be slightly higher (1.1 times) than the maximum VSS specific HAc-uptake rate of PAO II observed in this study. Only, when the HAc-uptake rate of GAO is higher than that of PAO at a high P-content, GAO may outcompete PAO, which apparently did not happen in this study under defined operational conditions (HAc, pH 7, 20 °C).

3.4.5 Implications on EBPR performance and modeling

The findings drawn in this study suggest that in wastewaters containing high VFA concentrations from hydrolysis and fermentation processes in the sewerage, the anaerobic kinetic rates of PAO are highly

Reference	(Suspected) organisms	SRT (d)	HRT (h)	P/C influent ratio (P-mol/C-mol)	Initial ISS/TSS (mg/mg)	Poly-P/active biomass[a] (P-mol/C-mol)	Pns/TSS[b] (mg/mg)	pH	Ca²⁺ (mg/L)	PHV/PHB (C-mol/C-mol)	PHV/HAc (C-mol/C-mol)	PHB/HAc (C-mol/C-mol)	P/HAc (P-mol/C-mol)	Gly/HAc (C-mol/C-mol)	qSA,ana^MAX [C-mmol/(gVSS.h)]	qP,ana^MAX [P-mol/(gVSS.h)]	qSA,ana^MAX [C-mmol/(C-mol.h)]	qP,ana^MAX [P-mmol/(C-mol.h)]
This study	PAO II	NA	NA	0	0.07	0.01	ND (0.02)	7 ± 0.1	3.8	0.27	0.32	1.22	0.01	1.01	2.3	NA	81	0
This study	PAO II	8	12	0.038	0.27	0.09	0.069	7 ± 0.05	3.8	0.19	0.26	1.35	0.19	1.22	3.5	0.9	121	31
This study	PAO II	8	12	0.038	0.29	0.11	0.072	7 ± 0.05	3.8	0.16	0.19	1.17	0.30	0.86	5.5	1.7	188	56
This study	PAO II	8	12	0.051	0.31	0.12	0.095	7 ± 0.05	3.8	0.15	0.16	1.07	0.33	1.02	6.0	1.9	198	64
This study	PAO II	8	12	0.077	0.40	0.16	0.130	7 ± 0.05	3.8	0.14	0.14	1.00	0.50	0.89	5.8	3.1	177	95
This study	PAO II	8	12	0.11	0.53	0.29	0.171	7 ± 0.05	3.8	0.12	0.12	0.97	0.61	0.66	4.9	3.2	157	103
Liu et al. (1997)	NA	8	6	0.078	NA	NA	0.13	7 ± 0.1	7.6	NA	NA	NA	0.66	0.42	6.0	3.5	NA	NA
Liu et al. (1997)	NA	8	6	0.038	NA	NA	0.08	7 ± 0.1	7.6	NA	NA	NA	0.46	0.65	3.5	1.5	NA	NA
Liu et al. (1997)	NA	8	6	0.008	NA	NA	0.02	7-8	7.6	NA	NA	NA	0.02	1.37	1.0	0.01	NA	NA
Schuler and Jenkins (2003a,b)	NA	4	12	0.003	NA	NA	0.018	7.15-7.25	16	NA	NA	NA	0.11	1.19	1.1[d]	NA	NA	NA
Schuler and Jenkins (2003a,b)	NA	4	12	0.006	NA	NA	0.036	7.15-7.25	16	NA	NA	NA	0.27	0.61	2.5[d]	NA	NA	NA
Schuler and Jenkins (2003a,b)	NA	4	12	0.012	NA	NA	0.051	7.15-7.25	16	NA	NA	NA	0.31	0.78	2.8[d]	NA	NA	NA
Schuler and Jenkins (2003a,b)	NA	4	12	0.019	NA	NA	0.082	7.15-7.25	16	NA	NA	NA	0.41	0.62	3.7[d]	NA	NA	NA
Schuler and Jenkins (2003a,b)	NA	4	12	0.039	NA	NA	0.13	7.15-7.25	16	NA	NA	NA	0.60	n.d	5.4[d]	NA	NA	NA
Schuler and Jenkins (2003a,b)	NA	4	12	0.062	NA	NA	0.16	7.15-7.25	16	NA	NA	NA	0.73	0.30	6.5[d]	NA	NA	NA
Schuler and Jenkins (2003a,b)	NA	4	12	0.085	NA	NA	0.14	7.15-7.25	16	NA	NA	NA	0.73	0.33	6.5[d]	NA	NA	NA
Schuler and Jenkins (2003a,b)	NA	4	12	0.105	NA	NA	0.16	7.15-7.25	16	NA	NA	NA	0.71	0.42	6.5[d]	NA	NA	NA
Welles et al. (2015)	PAO I	8	12	0.65	0.42	NA	0.14	7 ± 0.1	3.8	0.07	0.09	1.27	0.64	0.29	5.3	NA	179	NA
Welles et al. (2015)	PAO I	NA	NA	0.03	0.05	NA	ND (0.02)	7 ± 0.1	3.8	0.33	0.37	1.09	0.02	1.28	0.73	NA	23	NA
Welles et al. (2015)	PAO II	8	12	0.3	0.25	NA	0.068	7 ± 0.1	3.8	0.19	0.23	1.24	0.22	0.96	4.5	NA	154	NA
Welles et al. (2015)	PAO II	NA	NA	0.02	0.07	NA	ND (0.02)	7 ± 0.1	3.8	0.27	0.32	1.19	0.01	0.98	2.3	NA	80	NA
Zeng et al. (2003)	GAO	6.6	8	0.005	0.03	NA	NA	7 ± 0.1	6.8	0.38	0.52	1.39	NA	1.20	NA	NA	170	NA
Lopez-Vazquez et al. (2007)[c]	GAO	10	12	0.005	0.10	NA	NA	7 ± 0.1	3.8	0.1	0.69	1.28	0.01	1.20	NA	NA	200	NA

Table 3.2 Comparison of kinetic and stoichiometric values obtained in this study and previously reported parameters obtained with enriched EBPR cultures, and highly enriched PAO I, PAO II and GAO cultures.

NA = not applicable, ND = not detected (with the used FISH probes).

[a]: Calculated with equation 1 in material and methods

[b]: Calculated with equation 2 in material and methods

[c]: Stoichiometric values obtained from figures.

[d]: Calculated with equation in Figure 3 Schuler and Jenkins 2003b and P/HAc data from Table 3 Schuler and Jenkins 2003a

dependent on the poly-P content of PAO. In the anaerobic phase of WWTP's treating such wastewaters, PAO perform their best at a medium range P-content (like observed in this study on PAO II) or at a high P-content (when the biomass is presumably dominated by PAO I like in previous studies). However, as the P-content increases due to high influent P/C ratios, the return of phosphorus from the sludge line to the water line or if GAO prevail, the aerobic P-uptake ability of PAO is reduced because PAO tend to reach their saturation limit. In addition, the endogenous P-release activity of PAO increases, which may lead to higher P-release processes in an anaerobic zone that follow the aerated stage. Dependent on different factors such as the climate, the type of sewerage system and the dynamics of wastewater in the sewerage, the hydrolysis and fermentation processes in the sewerage and the associated VFA production, may be limited. In such cases, the PAO in the activated sludge systems may rely for a major extend on the VFA production from fermentation processes in the anaerobic stage of the activated sludge system, which are in general much slower than the VFA consumption processes by PAO I, II and GAO when fed with high VFA concentrations. In such cases, the competition between the different microbial communities will not be determined by the maximum rates of PAO I, II and GAO, but instead the competition between the organisms will be majorily be determined by the Ks values of the VFA uptake processes of the respective microbial communities.

The observed relationship between the anaerobic stoichiometry and the sludge P-content implies that a determination of the anaerobic HAc-uptake stoichiometry at pH 7.0 under strict anaerobic conditions at a temperature of 20ºC may help to estimate the poly-P content of the PAO present in the activated sludge. Without any microbial characterization, a stoichiometric P/HAc value in the range of 0.65-0.72 P-mol/C-mol in activated sludge at the end of the anaerobic phase, would indicate that the PAO present in the sludge are saturated with poly-P and that GAO are not present. In such cases, the net ortho-phosphate uptake capacity of the activated sludge may be limited due to incapability of PAO to take up additional phosphate during the anoxic or aerobic phase.

Considering the significant differences observed in the kinetic rates and stoichiometry of PAO at different P-contents and between the different PAO clades, it is recommendable to include the metabolic flexibility of the anaerobic metabolism in the current PAO model, as suggested by Acevedo et al. (2014), and to make specific metabolic PAO models for PAO I and PAO II. This would help to get a better description of EBPR performance in systems that work under dynamic conditions for the treatment of industrial wastewaters or combined biological and chemical P-removal processes.

3.5 conclusions

This study investigated the relationship between the storage polymers, the anaerobic kinetics rates and the stoichiometry of a highly enriched culture of 'Candidatus Accumulibacter phosphatis' Clade Type II. As the influent P/C ratio increased, the poly-P content of the biomass increased while its glycogen content decreased. At higher P-contents, the kinetic P-release rates for HAc-uptake and maintenance increased. In parallel, the HAc-uptake rates increased up to an optimal poly-P/glycogen ratio of 0.3 P-mol/C-mol. Above that optimal ratio, the HAc-uptake rate decreased. The stoichiometry of the anaerobic conversions showed that a metabolic shift occurred from a glycogen dependent metabolism towards a poly-P dependent metabolism when the poly-P content of the biomass increased. The changes in the HAc-uptake rates suggest that at low poly-P contents the ATP formation rate is the rate limiting step, while at high P-contents (and, thus, low glycogen contents) the NADH production rate becomes the rate limiting step for HAc-uptake. Electron microscopy showed that poly-P is stored in the form of large granules in each PAO cell and, therefore, the rate of poly-P consumption may be surface area limited. Therefore, a decrease in the poly-P content of the biomass could limit the ATP production and thereby trigger the ATP production from glycogen conversion at a smaller rate. The findings drawn from this study may help to understand the broad range of observed kinetic rates and stoichiometric values reported in previous studies and suggest that the metabolic flexibility of the specific PAO clades should be included in activated sludge models for a better description of EBPR processes.

3.6 Acknowledgements

This study was carried out as part of the SALINE project (http://www.salinesanitation.info) led by UNESCO-IHE Institute for Water Education and consortium partners KWR Watercycle Research Institute, Delft University of Technology, University of Cape Town, The Hong Kong University of Science and Technology, Polytechnic University José Antonio Echeverría and Birzeit University. Thanks to Mitchell Geleijnse who supported this study by conducting the phylogenetic analysis using the ppk1 gene.

3.7 References

Acevedo B., Oehmen A., Carvalho G., Seco A., Borras L., Barat R., (2012) Metabolic shift of polyphosphate-accumulating organisms with different levels of poly-phosphate storage. Water Res 46: 1889-1900

Acevedo B., Borras L., Oehmen A., Barat R., (2014) Modelling the metabolic shift of polyphosphate-accumulating organisms. Water Res 65: 235-244

Amann R.I. (1995) In situ identification of microorganisms by whole cell hybridization with rRNA-targeted nucleic acid probes. In: Akkermans ADL, van Elsas JD, de Bruijn FJ, editors. Molecular microbial ecology manual. London: Kluwer Academic Publisher, London, pp 1-15.

Barat, R., and van Loosdrecht, M.C.M. (2006) Potential phosphorus recovery in a WWTP with the BCFS process: Interactions with the biological process. Water Res 40(19): 3507-3516.

Barat, R., Montoya, T., Borras, L., Ferrer, J., and Seco, A. (2008) Interactions between calcium precipitation and the polyphosphate-accumulating bacteria metabolism. Water Res 42(13): 3415-3424

Bassin, J.P., Pronk, M., Muyzer, G., Kleerebezem, R., Dezotti, M., and van Loosdrecht, M.C.M. (2011) Effect of Elevated Salt Concentrations on the Aerobic Granular Sludge Process: Linking Microbial Activity with Microbial community Structure. Appl Environ Microbiol 77 (22): 7942-7953.

Braña A. F., Manzanal M. B., Hardisson C. (1980) Occurrence of polysaccharide granules in sporulating hyphae of Streptomyces viridochromogenes. Journal of bacteriology, 144(3), 1139-1142.

Brdjanovic D., van Loosdrecht M.C.M., Hooijmans C.M., Mino T., Alaerts G.J., Heijnen J.J., (1998) Effect of polyphosphate limitation on the anaerobic metabolism of phosphorus-accumulating microorganisms. Applied Microbiol Biotechnology 50, 273-276

Brdjanovic D., van Loosdrecht M.C.M., Hooijmans C.M., Alaerts G.J., Heijnen J.J. (1997) Temperature effects on physiology of biological phosphorus removal. J Environ Eng-ASCE 123(2): 144-154

Carvalho, G., Lemos, P.C., Oehmen, A., and Reis, M.A.M. (2007) Denitrifying phosphorus removal: linking the process performance with the microbial community structure. Water Res 41(19): 4383, 4396

Comeau, Y., Hall, K.J., Hancock, R.E.W., and Oldham, W.K. (1986) Biochemical-model for enhanced biological phosphorus removal. Water Res 20 (12): 1511–1521.

Crocetti G. R., Hugenholtz P., Bond P.L., Schuler A., Keller J., Jenkins D., Blackall L.L., (2000) Identification of polyphosphate-accumulating organisms and design of 16S rRNA-directed probes for their detection and quantitation. Appl Environ Microbiol 66: 1175-1182

Crocetti G.R., Banfield J.F., Keller J., Bond P.L., Blackall L.L. (2002) Glycogen accumulating organisms in laboratory-scale and full-scale wastewater treatment processes. Microbiol 148: 3353-3364.

Daims, H., Bruhl, A., Amann, R., Amann, R., Schleifer, K.H., and Wagner, M. (1999) The domain-specific probe EUB338 is insufficient for the detection of all bacteria: development and evaluation of a more comprehensive probe set. Syst Appl Microbiol 22: 345-352

Erdal U.G., Erdal Z.K., Daigger G.T., and Randall C.W., (2008) Is it PAO-GAO competition or metabolic shift in EBPR system? Evidence form an experimental study. Water Science and Technology 58(6), 1329-1334

Filipe C.D.M., Daigger G.T., Grady Jr. C.P.L. (2001) A metabolic model for acetate uptake under anaerobic conditions by glycogen accumulating organisms: Stoichiometry, Kinetics and the effect of pH.

Flowers J.J., He S., Yilmaz S., Noguera D.R., McMahon K.D. (2009) Denitrification capabilities of two biological phosphorus removal sludges dominated by different 'Candidatus Accumulibacter' clades. Environ Microbiol Rep (2009) 1(6): 583-588

Hesselman, R.P.X., von Rummell, R., Resnick S.M., Hany, R., Zehnder A.J.B. (2000) Anaerobic Metabolism of Bacteria Performing Enhanced Biological Phosphate Removal. Water Research., 34, 3487

Kamio Y., Terawaki Y., Nakajima T., Matsuda K. (1981) Structure of glycogen produced by Selenomonas ruminantium. Agricultural and Biological Chemistry, 45(1), 209-216.

Kisoglu, Z., Erdal, U., and Randall, C.W. (2000) The Effect of COD/TP ratio on Intracellular Storage Materials, System performance and Kinetic Parameters in a BNR system. Proceedings of the 73rd Annual Water Environment Federation Technical Exposition and Conference.

Kong Y.H., Beer M., Rees G.N. Seviour R.J. (2002) Functional analysis of microbial communities in aerobic-anaerobic sequencing batch reactors fed with different phosphorus/carbon)P/C) ratios. Microbiology 148, 2299-2307

Lanham A.B., Ricardo A.R., Coma M., Fradinho J., Carvalheira M., Oehmen A., Carvalho G., Reis M.A.M. (2012) Optiomisation of glycogen quantification in mixed microbial cultures. Biores. Technol.118, 518-525

Liebergesell M., Sonomoto K., Madkour M., Mayer F., Steinbüchel A. (1994) Purification and characterization of the poly (hydroxyalkanoic acid) synthase from Chromatium vinosum and localization of the enzyme at the surface of poly (hydroxyalkanoic acid) granules. European Journal of Biochemistry, 226(1), 71-80.

Liu W.T., Nakamura K., Matsuo T., Mino T. (1997) Internal energy-based competition between poly-phosphate- and glycogen-accumulating bacteria in biological phosphorus removal reactors-effect of P/C feeding ratio. Water Res 31(6): 1430-1438

Lopez-Vazquez CM, Song YI, Hooijmans CM, Brdjanovic D, Moussa MS, Gijzen HJ, van Loosdrecht MCM (2007) Short-term temperature effect on the anaerobic metabolism of glycogen accumulating organisms. Biotechnol Bioeng 97(3):483–495

Mayer, F., Madkour M. H., Pieper-Furst U., Wieczorek, R., Liebergesell, M., Steinbuchel, A. (1996) Electron microscopic observations on the macromolecular organization of the boundary layer of bacterial PHA inclusion bodies. The Journal of General and Applied Microbiology, 42(6), 445-455.

McMahon K. D., Yilmaz S., He S., Gall D. L., Jenkins D., Keasling J. D. 2007. Polyphosphate kinase genes from full-scale activated sludge plants. Appl Microbiol Biotechnol 77(1): 167-173.

Metcalf and Eddy, Inc. 2003. Wastewater Engineering - Treatment and Reuse, 4th Edn. Mc Graw Hill, New York, USA.

Mino, T., Arun, V., Tsuzuki, Y., and Matsuo, T., (1987) Effect of phosphorus accumulation on acetate metabolism in the biological phosphorus removal process. In Biological Phosphate Removal from Wastewaters, Advances in Water Pollution Control. Ramadori, R. (Ed). Oxford: Pergamon Press, pp. 27-38

Mino T., van Loosdrecht M.C.M., Heijnen J.J. (1998) Microbiology and Biochemistry of the Enhanced biological phosphate removal process. Water res. 32(11), 3193-3207.

Murnleitner E., Kuba T., van Loosdrecht M.C.M., Heijnen J.J. (1996) An integrated metabolic model for aerobic and denitrifying biological phosphorus removal. Biotechnol. Bioeng. 54(5), 434-450

Pereira, H., Lemos, P.C., Reis, M.A.M., Crespo, J.P.S.G. Carrondo, M.J.T., and Santos H. (1996) Model for Carbon Metabolism in Biological Phosphorus Removal Processes Based on In Vivo C13-NMR Labelling Experiments. Water Res 30: 2128

Pieper-Fürst U., Madkour M. H., Mayer F., Steinbüchel, A. (1994) Purification and characterization of a 14-kilodalton protein that is bound to the surface of polyhydroxyalkanoic acid granules in Rhodococcus ruber. Journal of bacteriology, 176(14), 4328-4337.

Satoh H., Mino T., Matsuo T. (1994) Deterioration of Enhanced Biological Phosphorus removal by the domination of microorganisms without polyphosphate accumulation. Wat. Sci. Tech. 30(6) 203-211

Schuler A.J., Jenkins D. (2003a) Enhanced Biological Phosphorus Removal from Wastewater by Biomass with Different Phosphorus Contents, Part 1: Experimental Results and Comparison with Metabolic Models. Water Environment Research 75(6), 485-498

Schuler A.J., Jenkins D. (2003b) Enhanced Biological Phosphorus Removal from Wastewater by Biomass with Different Phosphorus Contents, Part 2: Anaerobic Adenosine Triphosphate Utilization and Acetate Uptake Rates. Water Environment Research 75(6), 499-511

Skennerton C. T., Barr J. J., Slater F. R., Bond P. L., Tyson, G. W. (2014) Expanding our view of genomic diversity in Candidatus Accumulibacter clades. Environmental microbiology.

Smolders G.J.F., Van der Meij J., Van Loosdrecht M.C.M. and Heijnen J.J. (1994) Model of the anaerobic metabolism of the biological phosphorus removal process: Stoichiometry and pH influence. Biotechnol Bioeng 43: 461-470

Smolders, G.J.F., van Loosdrecht, M.C.M., and Heijnen, J.J. (1995) A Metabolic Model For the Biological Phosphorus Removal Process. Water Sci Technol (31): 79-97

Sudiana, I, Mino, T., Satoh, H., Nakamura, K., and Matsuo, T. (1999) Metabolism of Enhanced Biological Phosphorus Removal and Non-Enhanced Biological Phosphorus Removal Sludge with Acetate and Glycose as Carbon Source. Water Sci Technol (39): 29

Slater, F.R., Johnson, C.R., Blackall, L.L., Beiko, R.G., and Bond P.L. (2010) Monitoring associations between clade-level variation, overall community structure and ecosystem function in enhanced biological phosphorus removal (EBPR) systems using terminal-restriction fragment length polymorphism (T-RFLP). Water Res 44(17): 4908-4923

Steinbüchel A., Aerts K., Babel W., Föllner C., Liebergesell M., Madkour M.H., Mayer F., Pieper-Fürst U., Pries A., Valentin H. E., Wieczorek R. (1995) Considerations on the structure and biochemistry of bacterial polyhydroxyalkanoic acid inclusions. Canadian journal of microbiology, 41(13), 94-105.

Van Groenestijn J.W., Deinema M.H., Zehnder A.J.B. (1987) ATP production from polyphosphate in *Acinetobacter* strain 210A. Arch Microbiol 148, 14-19

Welles, L, Lopez-Vazquez, C.M., Hooijmans C.M., Van Loosdrecht, M.C.M., and Brdjanovic, D., (2014) Impact of salinity on the anaerobic metabolism of phosphate-accumulating organisms (PAO) and glycogen-accumulating organisms (GAO). Appl Microbiol Biotechnol 98(12): 7609-7622

Welles L., Lopez-Vazquez C. M., Hooijmans C. M., van Loosdrecht M. C. M., Brdjanovic, D. 2015. Impact of salinity on the aerobic metabolism of phosphate-accumulating organisms. Applied microbiology and biotechnology,99(8), 3659-3672

Welles L., Tian W. D., Saad S., Abbas B., Lopez-Vazquez C. M., Hooijmans C. M., van Loosdrecht M., Brdjanovic, D. 2015. Accumulibacter clades Type I and II performing kinetically different glycogen-accumulating organisms metabolisms for anaerobic substrate uptake. Water research, 83, 354-366.

Wentzel, M.C., Dold, P.L., Ekama, G.A., and Marais, G.v.R. (1985) Kinetics of biological phosphorus release. Water Sci Technol 17: 57-71

Wentzel, M.C., Dold, P.L., Loewenthal., R.E., Ekama, G.A., and Marais, G.v.R. (1987) Experiments towards Establishing the Kinetics of Biological Excess Phosphorus Removal. In Advances in Water Pollution Control: Biological Phosphate Removal from Wastewaters. Ramadori, R. (Ed). Oxford: Pergamon Press, pp. 79-91

Winkler M.K., Bassin J.P., Kleerebezem R., De Bruin L.M.M., Van den Brand T.P.H., van Loosdrecht M.C.M., 2011. Selective sludge removal in a segregated aerobic granular biomass system as a strategy to control PAOeGAO competition at high temperatures. Water Res. 45 (11), 3291-3299.

Zeng R.J., van Loosdrecht M.C.M., Yuan Z., Keller J. (2003) Metabolic model for glycogen-accumulating organisms in anaerobic/aerobic activated sludge systems. Biotechnol Bioeng 81(1): 92-105.

Zhou Y., Pijuan M., Zeng R.J., Lu H., Yuan Z., (2008) Could polyphosphate-accumulating organisms (PAO) be glyccogen-accumulating organisms (GAO)? Water Research 42, 2361-2368

3.8 Appendix 3A, Detailed description of equations for poly-P estimation

Poly-P was estimated on the basis of the ISS/TSS ratio and confirmed using steady-state mass balances as described in Welles et al. (2015). Equation 1 was developed using the ISS/TSS ratio of the biomass, assuming that (i) the ISS/TSS ratio associated with active biomass in non-EBPR biomass (ISS_b) was 0.025 mg ISS/mgTSS (as observed in this study after poly-P depletion), (ii) a poly-P composition of $(PO_3)_3MgK$ with a P-content ($f_{P,ppASH}$) of 0.31mgP/mgISS and (iii) negligible chemical precipitation. Equation 2, derived from the steady-state mass balance of phosphorus, assuming that (i) the solids in the effluent were negligible, (ii) a ratio of non poly-P phosphorus per VSS ($f_{P,bVSS}$) equal to the P-content of non-EBPR biomass at around 0.023 mgP/mgVSS (Metcalf and Eddy, 2003) and (iii) absence of chemical precipitation.

$ISS_{pp} = ISS - ISS_b$ (Eq. 1a)

$ISS_b = f_{ISSb,TSS,} / (1 - f_{ISSb,TSS}) * VSS$ (Eq. 1b)

$ISS_{pp} = ISS - f_{ISSb,TSS,} / (1 - f_{ISSb,TSS}) * VSS$ (Eq. 1c)

$poly\text{-}P = ISS_{pp} * f_{P,ppISS}$ (Eq 1d)

$poly\text{-}P = (ISS - f_{ISSb,TSS,} / (1 - f_{ISSb,TSS}) * VSS) * f_{P,ppISS}$ (Eq. 1e)

$Pns = poly\text{-}P + P_b$ (Eq 2a)

$P_b = f_{P,bVSS} * VSS$ (Eq 2b)

$d (T_{P,e} * V_p)/dt = Q_i * (T_{P,i} - T_{P,e}) - Q_w * TSS * f_{P,TSS} = 0$ (Eq 2c)

$Q_i / V_p * (T_{P,i} - T_{P,e}) - Q_w / V_p * TSS * f_{P,TSS} = 0$ (Eq 2d)

$1/HRT * (T_{P,i} - T_{P,e}) - 1/SRT * TSS * f_{P,TSS} = 0$ (Eq 2e)

$Pns = SRT/HRT * (T_{P,i} - T_{P,e})$ (Eq 2f)

$poly\text{-}P = SRT/HRT * (T_{P,i} - T_{P,e}) - f_{P,bVSS} * VSS$ (Eq 2g)

where;

TSS: Total suspended solids

VSS: Volatile suspended solids

ISS: Inorganic suspended solids

ISS_b: Inorganic suspended solids associated with active biomass

ISS_{pp}: Inorganic suspended solids associated with poly-P

Pns: Non-soluble total phosphorus

P_b: Phosphate associated with active biomass

Poly-P: Poly-phosphate

$T_{P,i}$: Total phosphorus concentration in the influent

$T_{P,e}$: Total phosphorus concentration in the effluent

$f_{P,TSS}$: Ratio of total P per TSS

$f_{P,bVSS}$: Ratio of non poly-P phosphorus per VSS

$f_{ISSb,TSS}$: ISS/TSS ratio associated with active biomass

$f_{P,ppISS}$: P-content of poly-P

V_p: Working volume of reactor

Q_i: Influent flow rate

Q_w: Wastage of activated sludge flow rate

HRT: Hydraulic retention time

SRT: Solids retention time

4

Prevalence of 'Candidatus Accumulibacter phosphatis' clade II under phosphate limiting conditions

Abstract

The competition between PAO and GAO cultures in EBPR processes has been the subject of several laboratory studies. Due to the current inability to grow these organisms in pure culture the research is relying fully on enrichment cultures of these organisms often using highly enriched PAO and GAO cultures. Phosphate limitation in EBPR systems has generally been used as a strategy to enrich GAO cultures. Recent studies have demonstrated that PAO are capable of performing a GAO metabolism when poly-P is absent and may be able to proliferate in phosphate limiting systems as well. Therefore, the objective of this study was to verify if phosphate limitation may also lead to mixed PAO-GAO or PAO cultures and, more specifically, which PAO clades may proliferate under phosphate limiting conditions. For this purpose, a sequencing batch reactor (SBR) was inoculated with activated sludge to enrich an EBPR culture for a cultivation period of 16 times the sludge age (128 days) under phosphate limiting conditions. A mixed PAO-GAO culture was obtained that comprised of 49% PAO II and 46% GAO. The mixed PAO-GAO culture was able to perform a typical GAO metabolism for the anaerobic uptake of HAc, but aerobically took up excessive amounts of phosphate. These findings suggest that the limitation of phosphate does not lead to the enrichment of only GAO. Therefore, microbial characterization is essential to confirm the actual cultures that dominate the enrichments. The prevalence of PAO II in this study supports the hypothesis that this PAO clade has a competitive advantage over PAO I under phosphate limiting conditions, suggesting that PAO II may be selected at low P/C influent ratios. This study also demonstrates that PAO are able to remain in activated sludge systems for periods of up to 16 SRT or longer when the influent phosphate concentrations are just enough for assimilation purposes and no poly-P is formed. They retain the ability to take up phosphate as soon as it becomes available in the influent.

Submitted as: Welles, L., Lopez-Vazquez, C. M., Hooijmans, C. M., van Loosdrecht M.C.M., Brdjanovic, D. (submitted). Prevalence of 'Candidatus Accumulibacter phosphatis' clade II under phosphate limiting conditions. *Applied Microbiology and Biotechnology*

4.1 Introduction

The enhanced biological phosphorus removal (EBPR) process is a cost-effective and environmental friendly process, often implemented in activated sludge systems for the removal of phosphate from wastewaters. Phosphate-accumulating organisms (PAO) are the organisms responsible for EBPR. These organisms take up VFA under anaerobic conditions and store it as PHA (Wentzel *et al.*, 1985, Comeau *et al.*, 1986, Mino *et al.*, 1987). The uptake and storage of VFA requires energy, which can be generated by poly-P cleavage and subsequent release of ortho-phosphate. In the following aerobic phase, PAO oxidize PHA and use the energy to restore their poly-P pool along with other metabolic processes. Another group of organism, the so called glycogen-accumulating organisms (GAO), are considered to compete with PAO for VFA, using a similar metabolism (Mino et al., 1987). The only difference is that GAO use glycogen instead of poly-P as energy source for the uptake and storage of HAc and therefore they do not contribute to enhanced biological P-removal. Prevalence of GAO is considered to be an important factor that leads to EBPR deterioration. Hence, the competition between PAO and GAO communitiesin EBPR processes has been the subject of several laboratory studies, often using highly enriched PAO and GAO cultures (Oehmen et al., 2007).

To study factors affecting PAO-GAO competition, it is necessary to obtain highly enriched PAO and GAO cultures. Influent phosphorus limitation has often been used as a strategy to enrich GAO (Sudiana et al., 1999; Lopez-Vazquez et al., 2007; Zeng et al., 2003; Filipe et al., 2001). Mostly, since it was considered that PAO were dependent on poly-P and would wash-out of the system once the phosphate concentrations became limiting. However, recent studies have shown that PAO are capable of performing a GAO metabolism (i.e. using glycogen to produce all of the required energy for VFA uptake) (Zhou et al., 2008; Acevedo et al., 2012). These new insights led to the speculation that P-limitation may not always lead to highly enriched GAO cultures. This would also imply that previous studies, thought to be conducted on enriched GAO cultures, may have actually been conducted on PAO cultures or mixed PAO-GAO cultures that performed a GAO metabolism if no microbial characterization methods were applied that could distinguish PAO and GAO populations(as in Sudiana *et al.*,1999; Liu *et al.*,1997; Schuler and Jenkins 2003b)). Therefore, it is important to (re-)investigate which microbial populations will develop in an EBPR systems operated under phosphate limiting conditions.

Besides the selection of PAO and GAO, it is also important to verify which PAO clades may proliferate under phosphate limiting conditions, as this may help to develop an enrichment strategy specifically for PAO I or II. The intrinsic differences within the metabolic characteristics of the PAO clades (Carvalho et al., 2007; Flowers et al., 2009; Slater et al., 2010), urges the need to study these metabolic differences and develop clade specific metabolic models for better understanding and description of EBPR processes. For this purpose, it is necessary to develop strategies for the selection of specific PAO clades to obtain EBPR cultures highly enriched with specific PAO clades.

A recent study suggested that PAO II may have a competitive advantage over PAO I under poly-P depleted conditions (Acevedo et al., 2012). Moreover, through short-term batch tests it was shown that the HAc-uptake rate of PAO II was four times faster than that of PAO I under poly-P depleted conditions (Welles et al., submitted). Therefore, low influent P/C ratios or the periodical phosphate limitation may lead to PAO II enrichments while high P/C influent ratios may support the development of PAO I cultures.

The objectives of this study were: (i) to reinvestigate which bacterial populations can develop in EBPR systems inoculated with activated sludge operated under phosphate limiting conditions, (ii) to assess whether PAO can get enriched and, if so, which PAO clade would prevail, and (iii) to verify whether PAO enriched under phosphate limiting conditions are capable of taking up excessive amounts of phosphate when they are suddenly exposed to high phosphate concentrations. This study contributes to the fundamental understanding of the PAO metabolism and clade differentiation, helps to improve the strategies for GAO and/or PAO II enrichment and providing insight into the EBPR performance of WWTPs that may (periodically) suffer from phosphate limiting conditions.

4.2 Material and Methods

4.2.1 Bacterial enrichment under phosphate limiting conditions

4.2.1.1 Operation of sequencing batch reactors (SBR)
An EBPR culture was enriched in a double-jacketed laboratory sequencing batch reactors (SBR) under the same conditions as those previously described for GAO by Welles et al. (2014). The SBR had a working volume of 2.5 L. Activated sludge from a municipal wastewater treatment plant (Hoek van Holland, The Netherlands) was used as inoculum. The SBR was operated in cycles of 6 hours (2.25h anaerobic, 2.25h aerobic and 1.5h settling phase) following similar operating conditions used in previous studies (Smolders *et al.*, 1994). The SBR was operated at a pH value of 7.0±0.05 and a temperature of 20±1°C. The applied total SRT was 8 days.

4.2.1.2 Synthetic medium
The SBR was fed with synthetic medium. The influent phosphate concentration was 2.2 mg PO_4^{3-}-P/L (0.07 P-mmol/L) (Liu et al., 1997) and the acetate concentration was 373 mg HAc/L (12.6 C-mmol/L, 400 mg COD/L, 860 mg NaAc·3H$_2$O), leading to an influent P/C ratio of 0.0056 (P-mol/C-mol). Further details regarding other macronutrients and trace elements can be found in Smolders et al. (1994).

4.2.1.3 Monitoring of SBR
The performance of the SBR was regularly monitored by measuring the mixed liquor suspended solids (MLSS) and mixed liquor volatile suspended solids (MLVSS). The (pseudo) steady-state conditions in the reactor was confirmed by daily determination of the aforementioned parameters as well as online pH and DO profiles.

4.2.2 Characterization of the microbial community
An estimation of the degree of enrichment of the bacterial populations of interest (PAO I, PAO II and GAO) was undertaken via fluorescence in situ hybridization (FISH) analyses. The whole bacterial community was targeted by the EUB338mix (general bacteria probe) (Amann *et al.*, 1990; Daims *et al.*, 1999). Betaproteobacteria and Gammaproteobacteria were identified with BET42 and GAM42a probes, respectively (Manz *et al.*, 1992). *Candidatus* Accumulibacter phosphatis was targeted by PAOMIX probe (mixture of probes PAO462, PAO651 and PAO846) (Crocetti *et al.*, 2000) whereas GAOMIX probe (mixture of probes GAOQ431 and GAOQ989) (Crocetti *et al.*, 2002) was used to target *Candidatus* Competibacter phosphatis. PAO I (clade IA and other type I clades) and PAO II (clade IIA, IIC and IID) were targeted by the probes Acc-1-444 and Acc-2-444 (Flowers *et al.*, 2009), respectively. Quantification of the PAO, GAO, PAO I and PAO II fractions was conducted by FISH quantification following the procedures described by Welles et al. (2014). The standard error of the mean (SEM) was calculated as the standard deviation of the area percentages divided by the square root of the number of images analysed.

4.2.3 Characterization of anaerobic carbon and phosphate conversions
When the SBR reached pseudo steady-state conditions, a cycle was intensively monitored to determine the biomass kinetic rates and stoichiometry of the anaerobic conversions. In addition to the above described parameters, orthophosphate (PO_4^{3-}-P), acetate (HAc-C), poly-beta-hydroxyalkanoate (PHA) and glycogen were measured in the cycle. To verify if the enriched culture was able to take up excessive amounts of

phosphate, two consecutive cycles were conducted in which a concentrated phosphate solution was added at the end of the anaerobic phase prior to the aerobic stage. This phosphate addition led to an increase of 22 mgP/L in the reactor. In the following aerobic phase, the phosphate concentration was intensively monitored.

4.2.4 Analyses

MLSS and MLVSS and PO_4^{3-}-P determination were performed in accordance with Standard Methods (A.P.H.A., 1995). HAc was determined using a Varian 430-GC Gas Chromatograph (GC), equipped with a split injector (split ratio 1:10), a WCOT Fused Silica column with a FFAP-CB coating (25 m x 0.53mm x 1μm), and coupled to a FID detector. Helium gas was used as carrier gas. The temperature of the injector, column and detector was 200°C, 105°C and 300°C, respectively. Glycogen was determined according to the method described by Smolders *et al.* (1994) but with an extended digestion of 5 h in 5 mL 0.9M HCl, using 5 mg of freeze dried biomass as described by Lanham et al. (2012). The PHB and PHV contents of freeze dried biomass were determined according to the method described by Smolders *et al.* (1994).

4.2.5 Determination of kinetic and stoichiometric parameters

The kinetic rate of interest was the anaerobic HAc-uptake rate. This rate was expressed as maximum active biomass specific rate based on the HAc profiles observed in the tests as described by Smolders *et al.* (1994b) and Brdjanovic et al., 1997). The stoichiometric ratios of interest were P/HAc, PHV/PHB, PHV/HAc, PHB/HAc and gly/HAc.

4.3 Results

4.3.1 Enrichment of EBPR culture under phosphate limiting conditions

After inoculation, the online pH and DO profiles indicated that within approximately 10 SRT (80 days) the performance of the reactor reached steady-state conditions. In Figure 4.1, the TSS, VSS and ISS/TSS concentrations are shown. Although the TSS and VSS concentrations fluctuated during the period of enrichment, the average TSS and VSS concentrations (1.9 gTSS/L, 1.8gVSS/L) of this period were in the range of concentrations observed in similar previous studies. The ISS/TSS ratio of the inoculum was 0.12 mg/mg, but it gradually decreased to a value of 0.04 mg/mg, except for the last data point which was obtained after conducting a batch experiment with an additional phosphate feed. The low ISS/TSS ratio observed under steady-state conditions indicates that the sludge did not contain poly-P and, if present, any chemical precipitates were negligible. As PAO are considered to rely on intracellular poly-P for anaerobic substrate uptake, it was expected that PAO were no longer present in this sludge.

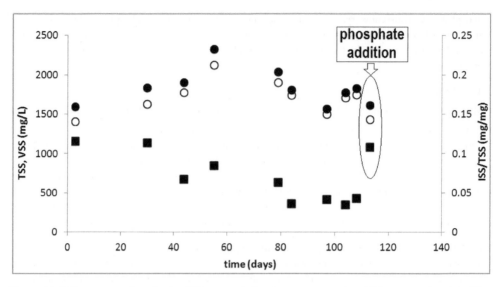

Figure 4.1 Solids concentrations in the SBR-reactor during long-term operation: TSS concentration (●), VSS concentration (○) and ISS/TSS ratio (■).

After approximately 14 SRT (113 days), the composition of the microbial community was analyzed. In Figure 4.2, FISH images and phase contrast images of the enriched culture are shown. Figure 4.2a shows remarkably that both PAO and GAO were present in similar quantities. Additional FISH analyses showed that among the PAO clades, PAO clade II was dominant whereas the presence of PAO clade I was negligible. FISH quantification showed that the PAO and GAO fractions of the total bacterial population were 49±6% and 46 ±7%, respectively. Based on these PAO and GAO fractions, the PAO/GAO ratio was around 1:1. Interestingly, the PAO I and II fractions with regard to the PAO population were 0±0% (n=16) and 99±2% (n=16), respectively. In Figure 4.2c and d, phase contrast images of suspected PAO and GAO cells are shown.

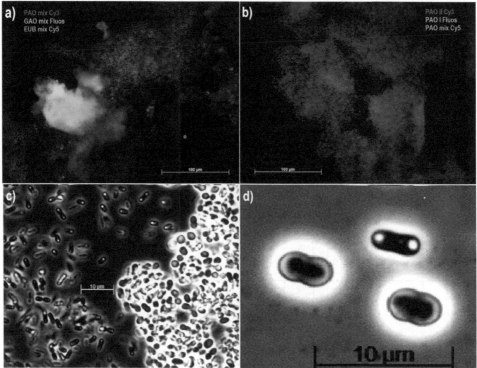

Figure 4.2 Representative FISH images (a,b) and phase contrast images (c,d) showing the distribution of bacterial populations in the enriched culture. In Figure (a); blue: all bacteria; purple: PAO; and cyan green: GAO. In Figure (b); blue: all PAO, purple: PAO clade II, and cyan green: PAO clade I. In Figure (c); dark cells: assumed to be PAO; bright: GAO. In Figure (d); small size: considered to be PAO; big size: considered to be GAO.

4.3.2 Biochemical conversions

To confirm the absence of a poly-P dependent metabolism for anaerobic substrate uptake under steady-state conditions, the carbon and phosphate conversions were monitored during one cycle after approximately 11 SRT (84 days) (Figure 4.3a). During the anaerobic phase, all HAc (6.25 C-mmol/L) was taken up in less than one hour with an active biomass specific HAc-uptake rate of 144 (C-mmol/(C-mol.h)) and this HAc uptake was coupled to a net P-release of 0.09 P-mmol/L (2.7 mgP/L). This led to a P/HAc ratio of 0.01 P-mol/C-mol, indicating a negligible involvement of a poly-P dependent metabolism. The stoichiometric values of all relevant anaerobic carbon and phosphate conversions are given in Table 4.1. The gly/HAc ratio, PHV/PHB as well as the other stoichiometric values are characteristic for a fully glycogen dependent metabolism as observed in enriched GAO cultures.

Reference	Reactor	Organisms	SRT (d)	HRT (h)	pH	Ca²⁺ (mg/L)	VSS/TSS (mg/mg)	PHV/PHB (C-mol/C-mol)	PHV/HAc (C-mol/C-mol)	PHB/HAc (C-mol/C-mol)	P/HAc (P-mol/C-mol)	Gly/HAc (C-mol/C-mol)	$q_{SA,ana}^{MAX}$ [C-mol/(C-mol.h)]
This study	SBR	PAO II and GAO	8	12	7+/-	3.8	0.96	0.37	0.54	1.45	0.03	1.28	0.14
Zeng et al. (2003)	SBR	GAO	6.6	8	7 +/- 0.1	6.8	0.97	0.38	0.52	1.39	NA	1.20	0.16-0.18
Lopez-Vasquez et al. (2007)[b]	SBR	GAO	10	12	7 +/- 0.1	3.8	0.9	0.34	0.69	1.28	0.01	1.20	0.20
Welles et al. (2015)	SBR	PAO II	8	12	7 +/- 0.05	3.8	0.75	0.19	0.23	1.24	0.22	0.96	0.15
Welles et al. (2015)	B	PAO II	NA	NA	7 +/- 0.1	3.8	0.93	0.27	0.32	1.19	0.01	0.98	0.08
Welles et al. (2015)	SBR	PAO I	8	12	7 +/- 0.1	3.8	0.58	0.07	0.09	1.27	0.64	0.29	0.18
Welles et al. (2015)	B	PAO I	NA	NA	7 +/- 0.1	3.8	0.95	0.33	0.37	1.09	0.02	1.28	0.02
Zhou et al. (2008)	SBR	PAO	8	24	7.0-8.0	1.3	0.6	0.06	0.07	1.18	0.62	0.46	NA
Zhou et al. (2008)	B	PAO	NA	NA	7.5+/-0.01	1.3	NA	0.37	0.46	1.24	0.06	1.03	0.07
Acevedo et al. (2012)	SBR	PAO I	8	12	7.0-8.9	10	0.45	0.04	0.05	1.31	0.7	0.38	NA
Acevedo et al. (2012)	SBR	PAO I,II	8	12	7.0-8.9	10	0.92	0.16	0.28	1.74	0.08	1.08	NA
Tian et al. (2013)	SBR	PAO I	16	12	7 +/- 0.1	3.8	NA	0.1	0.13	1.31	0.56	0.55	NA
Welles et al. (2014)	SBR	GAO	8	12	7.0	3.8	0.97	NA	NA	NA	0.012	1.2	0.15
Filipe et al. (2001)	SBR	?[b]	7	12	6.8-7.1	3.8	N.A.	0.31	0.38	1.26	N.A.	0.83	0.24
Sudiana et al. (1999)[a] reactor AL		?[c]	NA	NA	6.8-7.2	NA	NA	0.24	0.4	1.7	0.02	1.30	0.06-0.08
Liu et al. (1997)[a]	SBR	?	8	6	7 +/- 0.1	7.6	NA	NA	NA	NA	0.02	1.37	0.04
Schuler and Jenkins (2003b)[a]	SBR	?	4	12	7.15-7.25	16	NA	NA	NA	NA	0.11	1.19	0.03

Table 4.1 Stoichiometric values of the anaerobic conversions observed in this study and reported in previous studies with enriched PAO and GAO cultures.

[a]: calculated assuming that the VSS fully comprised of active biomass.

[b]: DGGE banding patterns indicated that 75% of the population belonged to γ-proteobacteria, while with FISH analysis, using specific probes developed to target the dominant γ-proteobacteria in the DGGE banding pattern, showed that 35% of the population stained positive for these γ-proteobacteria.

[c]: FISH analysis revealed that β-proteobacteria were dominant, comprising about one third of the sludge.

To verify if the PAO, that seemed to be present, were able to take up phosphate, two consecutive cycle tests were conducted after approximately 14 SRT (113 days), in which 0.8 P-mmol/L (25 mgP/L) was added to the reactor at the end of each anaerobic phase (Figure 4.3b). In the first cycle, 0.6 P-mmol/L (18 mgP/L) was taken up while in the second cycle 0.7 P-mmol/L (21 mgP/L) was taken up (data not shown). This confirms that the PAO observed by microscopy were present and active, and that they were able to remove excessive amounts of phosphate from the liquid phase. This was further confirmed by the increase in the ISS/TSS ratio. The storage of 39 mgP/L as poly-P is equivalent to 0.13 gISS/L, assuming a poly-P composition of (KMg(PO₃)₃). With a TSS concentration of 1.6 gTSS/L, this increase in the ISS concentration would make a difference of 0.08 mg/mg in the ISS/TSS ratio, which approximately corresponds to the increase in the ISS/TSS ratio shown in Figure 4.1.

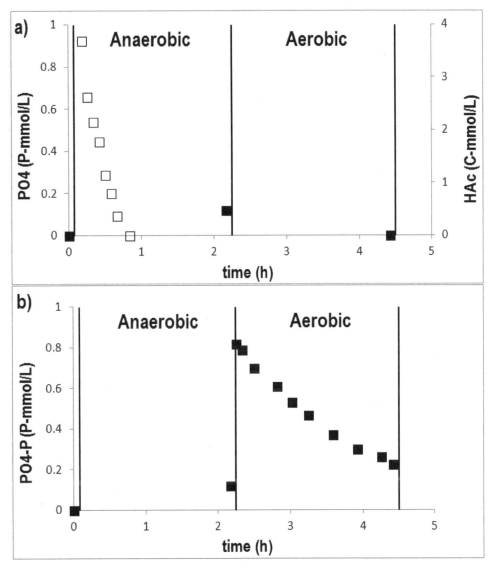

Figure 4.3 Acetate (Hac, □) and ortho-phosphate (PO₄, ■) conversions observed during: (a) a steady state cycle and (b) after the addition of phosphate in the end of the anaerobic phase.

4.4 Discussion

4.4.1 Enrichment of mixed PAO-GAO cultures

Bacterial populations cultivated in continuous and sequencing batch reactors will always wash-out (i) from continuously operated reactors if the maximum growth rate of the organisms is lower than the dilution rate or (ii) in SBR reactors if the maximum net growth per cycle is lower than the net biomass removal per cycle. If, bacteria are not able to grow at all then in the ideal caseit would take only 1 SRT to wash-out more than 64% of the bacterial population that was present originally and 3 SRT to wash-out at least 95%. In this study a mixed PAO-GAO culture was obtained after an enrichment period of 14 SRT under phosphate limiting conditions, which comprised 36% PAO II and 32% GAO. Therefore, this study clearly demonstrated that PAO II were able to proliferate under phosphate limiting conditions. In a previous study, the PAO fractions were determined in six activated sludge treatment plants in the Netherlands (Lopez-Vazquez et al., 2008a), ranging from 6% to 16 % (with an average of 9.2%) of the total bacterial community, while the GAO fractions ranged from 0.4 to 3.2% (showing an average of 1.7%). The considerably higher PAO fraction (36%) obtained in the enriched culture of this study suggests that with practically depleted poly-P reservesPAO II were still able to compete with ordinary heterothropic bacteria by taking up VFA in the anaerobic phase using a metabolism that is not dependent on poly-P. The stoichiometry of the anaerobic carbon and phosphate conversions confirmed that the mixed population made full use of a GAO metabolism under anaerobic conditions (Table 4.1). Additional experiments confirmed that PAO were able to perform a PAO metabolism when phosphate was added to the system (Figure 4.3). These results are in line with previous findings regarding a GAO enrichment study performed by Lopez-Vazquez et al. (2007), in which a mixed culture was obtained containing 75% GAO and 20% PAO after an enrichment period of 10 SRT. In two previous GAO enrichment trials performed by the authors of the present study, a reactor was inoculated with sludge from a highly enriched PAO II reactor (Welles et al., 2015) assuming that the sludge contained GAO since most of HAc was taken up through a GAO metabolism. In addition, it was expected that any PAO present would rapidly be washed out of the system due to the low influent P concentration (assuming that they were fully dependent on poly-P) leading to a highly enriched GAO culture. However, in those enrichment trials FISH analyses revealed that (i) the sludge used as inoculum was rich in PAO II and did not contain GAO (Welles et al., 2015) and (ii) that under phosphate limiting conditions (like those applied in the present study) PAO II were still dominant after 15 and 16 SRT of operation (data not shown). In spite of the presumably low poly-P contents, those cultures were able to completely remove HAc during the anaerobic phase.

4.4.2 Competition between PAO I, PAO II and GAO

Enrichment of the specific PAO clade II is in agreement with the findings of Welles et al. (submitted), where PAO II showed HAc uptake rates four times higher than those of PAO I under poly-P depleted conditions. In those studies, it was suggested that PAO II has a competitive advantage over PAO I under phosphate limiting conditions. In a study performed by Tian et al. (2013), an enriched PAO I culture was not able to complete the HAc-uptake during the anaerobic phase (36d SRT, 0.5d HRT, at 10°C, pH 7.0 with an influent HAc concentration of 375 mg/L and an anaerobic phase of 2.25h) when the phosphate concentration was limiting (2.2mgP/L influent ortho-phosphate concentration, resulting in an influent P/C ratio of 0.0056 P-mol/C-mol). Similar to this observation, Schuler and Jenkins (2003) also observed a leakage of HAc to the aerobic phase (4d SRT, 0.5d HRT, 20°C, pH 7.0, 188 mg/L influent HAc concentration and 1.83h anaerobic phase duration) when phosphate became limiting (at an influent P/C ratio lower than 0.019 P-mol/C-mol). Based on the stoichiometry and kinetic rates reported, it seemed that the culture of Schuler and Jenkins (2003) was also a PAO I dominated culture (Welles et al., submitted). The findings in this study, Tian et al. (2013) and Schuler and Jenkins (2003) support the hypothesis that under P-limiting conditions, PAO II can proliferate in the system by adjusting its metabolism to low P/C influent ratios while PAO I cannot. At high influent P/C ratios (above 0.04 P-mol/C-mol) with a high poly-P content, PAO I may exhibit faster HAc-uptake rates and outcompete PAO II.

4.4.3 PAO II and GAO

In the study of Welles et al. (submitted), the active biomass specific HAc-uptake rates of PAO (80 C-mmol/(C-mol.h)) determined in short-term batch tests under poly-P depleted conditions seemed to be significantly lower than that of GAO (150-200 C-mmol/(C-mol.h) (Zeng et al., 2003, Lopez-Vazquez et al., 2007; Welles et al., 2014), suggesting that GAO would still be able to outcompete PAO after a few SRT. However, the PAO fractions observed even after enrichment periods of 14-16 SRT were still very significant. This indicates that a population shift from PAO II to GAO needed a long period due to a high initial PAOII/GAO ratio and a relatively small difference in the HAc-uptake rates of the PAO and GAO enriched in this system. Alternativeley, GAO do not have determinant competitive advantages over PAO II under poly-P limiting conditions for instance due to potential differences in the biomass yield of PAO and GAO, which could affect the PAO-GAO competition as well.

4.4.4 Implications

Limitation of phosphorus has often been used as a strategy to enrich GAO (Sudiana *et al.*, 1999; Filipe *et al.*, 2001; Zeng et al.,2003 and Lopez-Vazquez et al., 2007). This study has demonstrated that it is an unreliable selection strategy that does not necessarily lead to highly enriched GAO cultures. In addition, the carbon and phosphorus conversions cannot be used as reliable indicators to assess the presence of GAO. Therefore, to conduct a microbial characterization, using FISH or other microbial identification techniques is always recommended in EBPR studies. Results of previous GAO studies performed without any microbial identification (e.g. Sudiana *et al.*,1999; Liu *et al.*,1997; Schuler and Jenkins 2003b)) may therefore be questioned, especially if the enrichment period was limited to few SRT only. The observation that both PAO and GAO are able to perform a GAO metabolism but at different rates helps to explain the broad range of HAc-uptake rates reported from enriched 'GAO' cultures. In the studies of Sudiana *et al.* (1999), Liu *et al.* (1997) and Schuler and Jenkins (2003b), where the presence of GAO was not reported, HAc-uptake rates range from 0.04 to 0.08 C-mol/(C-mol biomass.h), while the HAc-uptake rates in studies where the presence of GAO was confirmed vary between 0.16-0.20 C-mol/(C-mol biomass.h) (Lopez-Vazquez et al., 2007; Zeng et al., 2003; Filipe *et al.*, 2001; Welles et al. 2014). Possibly under P-limited conditions lower HAc uptake rates could be associated to PAO II enrichments and higher rates to GAO enrichments.

This study also suggest that the PAO II/GAO fractions in the inoculum have a significant impact on the time that is needed for enrichment of GAO cultures and that a long enrichment period may be needed to obtain highly enriched cultures. Welles et al. (2014) were able to obtain a highly enriched GAO culture after 44 SRT (352 days) under the same operational conditions, although minor traces of PAO II were still present in the biomass. The selection of GAO may be accelerated when the operating temperature of the GAO is increased (Lopez-Vazquez et al., 2007, 2008), but on the other hand this could lead to selection of specific GAO that normally prevail in processes at elevated temperature and therefore those GAO cultures may not be representative in context of the research on wastewater treatment in moderate climate conditions. Further research is needed to define appropriate strategies to enrich PAO I, PAO II and GAO for their further study and characterization.

The findings drawn from this study also indicate that temporary limitation of ortho-phosphate by temporal overdosing of iron in activated sludge systems or fluctuations in the influent P/C ratio of industrial wastewater may not have deleterious effects on the ability of the activated sludge to perform EBPR once the ortho-phosphate levels in the influent are restored. Past studies have indicated that simultaneous chemical precipitation and enhanced biological phosphorus removal in activated sludge systems, led to a decreased biological phosphorus removal activity and accompanied storage of Poly-P during the periods of iron addition, due to a competition for ortho-phosphate by the chemical and biological mechanisms (de Haas et al., 2000; de Haas et al., 2004). However, depletion of poly-P storage pools in PAO II, would not severaly affect its ability to proliferate in the system and thereforePAO II can remain in the systems for several SRT while performing a GAO metabolism. Once the ortho-phosphate concentrations are restored, PAO II may be still present in the system and perform EBPR activity instantly. Temporal overdosing of iron in activated sludge systems or fluctuations in the influent P/C ratio of industrial wastewaters may be problematic in activated sludge systems dominated by PAO I but not in sludge systems dominated by PAO II. Based on the duration of this study and considering that the applied SRT in full-scale activated sludge plants achieving phosphorus and nitrogen removal usually vary between 8 to 30 days, PAO II could be able to prevail in activated sludge systems for 80 up to 300 days under phosphate limiting conditions.

4.5 Conclusions

A mixed PAO-GAO culture was enriched after a cultivation period of 14-16 SRT under ortho-phosphate limiting conditions. The PAO and GAO fractions of the total microbial community were around 49% and 46%, respectively. In particular, all PAO were closely related to *Candidatus* Accumulibacter phosphatis Clade II. Under anaerobic conditions, the mixed PAO-GAO culture performed a typical GAO metabolism in which all energy for HAc-uptake was produced by the conversion of glycogen. However, under aerobic conditions PAO were capable of taking up excessive amounts of phosphate when additional phosphate was added to the reactor. This study suggests that limitation of phosphate, often used as a strategy for the enrichment of GAO, does not necessarily lead to high GAO enrichment and that the carbon conversions often used as indicator for GAO enrichments are no longer reliable as stand-alone indicators. Furthermore, the development of PAO II suggests that PAO II have a competitive advantage over PAO I under phosphate limiting conditions. From a practical perspective, this study demonstrates that PAO may be able to proliferate under phosphate limiting conditions in activated sludge systems for periods of up to 16 SRT or longer while being able to take up phosphate aerobically as soon as it is available in the influent.

4.6 Acknowledgements

This study was carried out as part of the SALINE project (http://www.salinesanitation.info) led by UNESCO-IHE Institute for Water Education with the consortium partners KWR Watercycle Research Institute, Delft University of Technology, University of Cape Town, The Hong Kong University of Science and Technology, the Higher Polytechnic Institute "José Antonio Echeverría" and Birzeit University.

4.7 References

Acevedo B., Oehmen A., Carvalho G., Seco A., Borras L., Barat R. (2012) Metabolic shift of polyphosphate-accumulating organisms with different levels of poly-phosphate storage. Water Research 46, 1889-1900

Carvalho G., Lemos P.C., Oehmen A., Reis M.A.M. (2007) Denitrifying phosphorus removal: linking the process performance with the microbial community structure. Water Research 41(19), 4383, 4396

Comeau Y., Hall K.J., Hancock R.E.W., Oldham W.K. (1986) Biochemical-model for enhanced biological phosphorus removal. Water Res. 20 (12), 1511–1521.

De Haas, D. W., Wentzel, M. C., & Ekama, G. A. (2000). The use of simultaneous chemical precipitation in modified activated sludge systems exhibiting biological excess phosphate removal. Part 4: Experimental periods using ferric chloride. Water Sa, 26(4), 485-504.

De Haas, D. W., Wentzel, M. C., & Ekama, G. A. (2004). The use of simultaneous chemical precipitation in modified activated sludge systems exhibiting biological excess phosphate removal: Part 5: Experimental periods using a ferrous-ferric chloride blend. Water Sa, 27(2), 117-134.

Filipe CDM, Daigger GT, Grady Jr CPL (2001) A metabolic model for acetate uptake under anaerobic conditions by glycogen-accumulating organisms: stoichiometry, kinetics and effect of pH. Biotechnol Bioeng 76(1):17-31.

Flowers J.J., He S., Yilmaz S., Noguera D.R., McMahon K.D. (2009) Denitrification capabilities of two biological phosphorus removal sludges dominated by different 'Candidatus Accumulibacter' clades. Environmental Microbiology Reports (2009) 1(6), 583-588

Liu WT, Nakamura K, Matsuo T, Mino T (1997) Internal energy-based competition between polyphosphate- and glycogen-accumulating bacteria in biological phosphorus removal reactors-effect of P/C feeding ratio. Water Res 31(6):1430-1438.

Lopez-Vazquez, C.M., Song YI, Hooijmans, C.M., Brdjanovic, D., Moussa MS, Gijzen, H.J., Van Loosdrecht, M.C.M., (2007) Short-term temperature effects on the anaerobic metabolism of Glycogen Accumulating Organisms. Biotechnol. Bioeng. 97(3):483-495

Lopez-Vazquez, C.M., Hooijmans, C.M., Brdjanovic, D., Gijzen, H.J., Van Loosdrecht, M.C.M. (2008a) Factors affecting the microbial populations at full-scale enhanced biological phosphorus removal (EBPR) wastewater treatment plants in The Netherlands. Water Research 42(10), 2349-2360

Lopez-Vazquez CM, Song YI, Hooijmans CM, Brdjanovic D, Moussa MS, Gijzen HJ, van Loosdrecht MCM (2008) Temperature effects on the aerobic metabolism of glycogen accumulating organisms. Biotech Bioeng 101(2):295-306.

Mino T., Arun V., Tsuzuki Y., Matsuo T., (1987) Effect of phosphorus accumulation on acetate metabolism in the biological phosphorus removal process. In: Ramadori, R. (Ed.), Biological Phosphate Removal from Wastewaters, Advances in Water Pollution Control, vol 4. Pergamon Press, Oxford (1987), pp. 27-38

Oehmen A, Lemos PC, Carvalho G, Yuan Z, Keller J, Blackall LL, Reis MAM. (2007) Advances in enhanced biological phosphorus removal: From micro to macro scale. Wat Res 41(11):2271-2300.

Welles, L; Lopez-Vazquez, C.M., Hooijmans C.M., Van Loosdrecht, M.C.M. Brdjanovic, D., (2014) Impact of salinity on the anaerobic metabolism of phosphate -accumulating organisms (PAO) and glycogen -accumulating organisms (GAO). Appl. Microbiol. biotechnol. 98(12), 7609-7622

Welles, L., Lopez-Vazquez, C. M., Hooijmans, C. M., van Loosdrecht, M. C. M., & Brdjanovic, D. (2015). Impact of salinity on the aerobic metabolism of phosphate-accumulating organisms. Applied microbiology and biotechnology,99(8), 3659-3672. Welles L., Abas B., Lopez-Vazquez C. M., Hooijmans C.M., van Loosdrecht M.C.M. and Brdjanovic. D. (submitted) Effect of storage polymers on the kinetics and stoichiometry of Candidatus Accumulibacter phosphatis clade II. Applied and Environmental Microbiology

Welles, L., Tian, W. D., Saad, S., Abbas, B., Lopez-Vazquez, C. M., Hooijmans, C. M., ... & Brdjanovic, D. (2015). Accumulibacter clades Type I and II performing kinetically different glycogen-accumulating organisms metabolisms for anaerobic substrate uptake. Water research, 83, 354-366.

Schuler A.J., Jenkins D. (2003a) Enahnced Biological Phosphorus Removal from Wastewater by Biomass with Different Phosphorus Contents, Part 1: Experimental Results and Comparison with Metabolic Models. Water Environment Research 75(6), 485-498

Schuler A.J., Jenkins D. (2003[b]) Enahnced Biological Phosphorus Removal from Wastewater by Biomass with Different Phosphorus Contents, Part 2: Anaerobic Adenonsine Triphosphate Utilization and Acetate Uptake Rates. Water Environment Research 75(6), 499-511

Slater F.R., Johnson C.R., Blackall L.L., Beiko R.G., Bond P.L. (2010) Monitoring associations between clade-level variation, overall community structure and ecosystem function in enhanced biological phosphorus removal (EBPR) systems using terminal-restriction fragment length polymorphism (T-RFLP). Water Res 44(17), 4908-4923

Sudiana IM, Mino T, Satoh H, Nakamura K, Matsuo T (1999) Metabolism of enhanced biological phosphorus removal and non-enhanced biological phosphorus removal sludge with acetate and glucose as carbon source. Wat Sci Tech 39(6):29-35.

Tian, W.D., Lopez-Vazquez, Li, W.G., C.M., Brdjanovic, Van Loosdrecht, M.C.M. (2013) Occurrence of PAOI in a low temperature EBPR system. Chemosphere 92, 1314-1320

Wentzel M.C., Dold P.L.., Ekama G.A. and Marais G.v.R. (1985) Kinetics of biological phosphorus release. Water Science and Technology 17, 57-71

Zeng RJ, van Loosdrecht MCM, Yuan Z, Keller J (2003) Metabolic model for glycogen-accumulating organisms in anaerobic/aerobic activated sludge systems. Biotechnol Bioeng 81(1):92–105

Zhou Y., Pijuan M., Zeng R.J., Lu H., Yuan Z., (2008) Could polyphosphate-accumulating organisms (PAO) be glyccogen-accumulating organisms (GAO)? Water Research 42, 2361-2368.

5

Denitrification pathways of PAO clade I with different carbon sources

Abstract

Past studies indicated that PAO are capable of using nitrate as external electron acceptor. This allows for an efficient integration of nitrogen and phosphate removal with minimal need for organic carbon (COD). In different engineering processes and system configurations that have been developed, contradicting findings appeared regarding the denitrification capacities of PAO. Whereas some studies suggested that only PAO clade I might be capable of using nitrate as external electron acceptor for anoxic P-uptake, other studies suggested that PAO II may be responsible for anoxic P-removal, but the findings in past studies are inconclusive due to the lack of EBPR PAO cultures highly enriched with specific PAO clades. In the present study, a PAO I culture was enriched (>99%) in an SBR operated under anaerobic/oxic conditions in a first phase and exposed to anaerobic/anoxic/oxic conditions in a second phase to acclimatize the sludge to the presence of nitrate. In the first phase (before acclimatization) and in the second phase (after acclimatization), the aerobic and anoxic (nitrate and nitrite) activities were assessed in batch tests when biomass was fed anaerobically with either acetate or propionate. When nitrate was the electron acceptor, no P-uptake was observed before and even after acclimatization. Only limited nitrate removal was observed at a low rate (0.1 mmol NO_3-N/(L.h)), suggesting that this removal may have occurred due to the activity of flanking bacteria. With nitrite, simultaneous P and nitrite removal was observed after acclimatization. Before acclimatization, nitrite removal rates were in particular lower in the acetate fed reactor and no phosphate uptake occurred while in the propionate fed reactor some phosphate uptake occurred. This study suggests that PAO I is only able to use nitrite as external electron acceptor under Anaerobic/Anoxic/Oxic conditions. The findings therefore suggest that formation of nitrite by bacteria other than PAO I in Anaerobic/Anoxic/Oxic systems is essential for anoxic phosphate removal by denitrifying PAO.

Submitted as:

Saad S., Welles L., Abbas B., Lopez-Vazquez M.C, van Loosdrecht M.C.M., Brdjanovic D. (submitted) Denitrification pathways of PAO clade I with different carbon sources. *Water Research*

5.1 Introduction

In the enhanced biological phosphorus removal (EBPR) process, sludge is cycled through anaerobic and aerobic/anoxic zones (Mino et al., 1998; Oehmen et al., 2007). The group of microorganisms that is responsible for the EBPR performance is broadly known as Polyphosphate-Accumulating Organisms (PAO). These organisms are able to anaerobically take up volatile fatty acids and store them in the form of PHAs, and acquire energy from the degradation of the stored phosphate as intracellular polyphosphate, releasing ortho-phosphate to the water bulk phase. In the following aerobic phase PAO grow and take-up ortho-phosphate to recover their poly-P pools, leading to P removal from the bulk liquid via PAO cell removal by wastage of activated sludge (Mino et al., 1987; Henze et al., 1997; Mino et al., 1998; Tchobanoglous et al., 2008)

Several studies demonstrated that PAO can also grow and take up phosphate under anoxic conditions. The so-called denitrifying polyphosphate-accumulating organisms (DPAO) are thought to be capable of using both nitrate and oxygen as electron acceptors instead of solely oxygen (Kerrn-Jespersen and Henze, 1993; Kuba et al., 1993; Hu et al., 2002), which allowed the development of processes for combined nitrogen and phosphate removal with minimal COD consumption.

Different engineering processes and system configurations were developed to achieve simultaneous biological nitrogen and phosphorus removal with the help of DPAO. The Modified University of Cape Town (MUCT) process is generally used for EBPR processes with optimized nutrient removal (Henze et al., 2008). The cycling between anaerobic and anoxic zones in this process is believed to sustain anoxic P-removal. Some processes have been proposed to optimize the removal of phosphate by DPAO, the best studied is the Dephanox (or A2/N) process, where nitrification sludge is separated from the P-removal/denitrification sludge (Bortone et al., 1996; Kuba et al., 1996).

However, contradicting results have been reported in literature. In some studies, it was observed that anoxic P-uptake gradually stopped while nitrate was still available. In these studies, P-uptake continued in the subsequent aerobic phase which was thought to be related to depleted PHA pools of DPAO under anoxic conditions (Parco et al., 2007), while normal PAO (not capable of using nitrate) still contained PHA and came active under aerobic conditions (Kerrn-Jespersen and Henze 1993; Meinhold et al., 1999). This observation led to the distinction of two types of PAO with different denitrification capabilities namely PAO that were only capable of using oxygen as external electron acceptor and DPAO that were able to use nitrate and nitrite as external electron acceptor. On the contrary, in several following studies it was suggested that there are no differences between PAO and DPAO (Ahn et al., 2002; Kong et al., 2002; Zeng et al., 2003b).

In the first studies that focused on the differentiation of the PAO clades responsible for denitrification, DPAO and PAO were distinguished by morphology, where rod type morphology was observed to be dominant in cultures that succeeded to denitrify via nitrate pathway, while cocci type was more abundant in the ones that failed (Carvalho et al., 2007). Based on FISH analysis, the full denitrification capacity (starting from nitrate) of PAO was attributed to PAO I species (DPAO) rather than PAO II which was thought to be only capable of using oxygen and nitrite (Carvalho et al., 2007; Flowers et al., 2009; Oehmen et al., 2010a; Lanham et al., 2011).

The presence of propionate as a more reduced carbon source has been suggested to lead to conditions that require a high energy demand relative to the demand of reduction equivalents which may favor in general PAO over GAO, where several researchers reported that propionate is more favorable for EBPR than acetate (Thomas et al., 2003; Pijuan et al., 2004; Oehmen et al., 2005a; Carvalho et al., 2007;

Broughton et al., 2008; Li et al., 2008; Vargas et al., 2011). The presence of propionate was accompanied by the dominance of PAO I in some studies (Carvalho et al., 2007; Lanham et al., 2011). However, (Gonzalez-Gil and Holliger, 2011; Guerrero et al., 2012) fed the system with propionate and they found PAO IIA to be dominant.

Several studies reported that PAO I was not able to perform denitrification of nitrate due to the lack of the nitrate reductase, and it was suggested that PAO I rely on other flanking microorganisms or denitrifying GAO for the nitrite supply under anoxic conditions (Zeng et al., 2003a; Martin et al., 2006; Guisasola et al., 2009; Kim et al., 2013). Furthermore, in the studies from (Guerrero et al., 2012; Wang et al., 2014) with enriched PAO cultures dominated by PAO II (78% and 54% respectively), denitrification of nitrate was also observed.

The significant presence of microbial populations other than PAO in many past studies and lack of cultures highly enriched with only one specific type of PAO, have hampered the elucidation of functional differences among PAO clades regarding their denitrification capabilities. For instance, (Flowers et al., 2009) conducted in a study with two cultures containing different ratios of PAO clades (culture 1: 67% PAO I and 5% PAO II 28% other ; culture two 32% PAO I and 50 % PAO II, 18% other). Both cultures showed similar initial uptake rates of nitrate, but the nitrate uptake rate of the PAO II dominated culture started to level off in a shorter period of time when compared to the PAO I dominated culture. Flowers et al., (2009) suggested that the PAO I in both cultures were responsible for the initial denitrification rates and that the rates started to level off in a shorter period of time in the PAO II dominated culture as the PHA content of the smaller PAO I fraction got depleted in an earlier stage. However, when the biomass specific rates are determined for the specific PAO clades, the findings contradict their hypothesis. Furthermore, the differences in denitrification rates in each system can also be explained due to difference in abundance of side populations that may have been (partially) responsible for the denitrification activity.

Regarding the acclimatization period to nitrate, the study by (Zeng et al., 2003b) suggested that only few hours were needed for the acclimatization of PAO to nitrate in order to obtain the denitrifying activity. Furthermore, several studies indicated that long acclimatization periods, which were considered to be essential for obtaining denitrification capabilities, led to a change in the microbial populations, decreasing the PAO fractions present in the biomass cultures (Carvalho et al., 2007; Flowers et al., 2009; Guisasola et al., 2009).

Overall, the results of previous studies are contradictory and inconclusive as many studies lack information on the prevailing microbial populations or were not conducted with cultures that were highly enriched with specific PAO.

The objective of this study was to assess the anoxic (NO_2- and NO_3-) activity of a highly enriched PAO I culture in comparison to the aerobic activity before and after acclimatization to nitrate after feeding with either acetate or propionate. By opting at a short acclimatization period, the biomass had a chance for adaptation, but the potential change in the microbial community was minimal, and different carbon sources were used to verify if the carbon source also affects the ability of PAO to denitrify.

5.2 Material and Methods

5.2.1 Enrichment of PAO

5.2.1.1 Operation of Parent SBR

PAO were enriched and cultivated during two different experimental phases. In phase I, a PAO culture was enriched and cultivated under Anaerobic/Oxic conditions in a double-jacketed laboratory sequencing batch reactor (SBR), while in phase II PAO were cultivated under Anaerobic/Anoxic/Oxic conditions. The SBR was operated and controlled automatically by an Applikon ADI controller. Online operating data (e.g. pH and O_2) was stored using BioXpert software (Applikon, The Netherlands, Schiedam). The reactor had a working volume of 2.5 L. Activated sludge from Harnaschpolder WWTP a municipal wastewater treatment plant in The Netherlands, was used as inoculum for the enrichment of the PAO culture.

5.2.1.2 Operation in anaerobic/oxic mode

In the first operation mode, the SBR was operated in cycles of 6 hours (135 min anaerobic, 135 min aerobic and 90 min settling and decanting phase) following similar operating conditions used in previous studies (Smolders et al., 1994; Brdjanovic et al., 1997). pH was maintained at 7.6 ± 0.05 by dosing 0.4 M HCl and 0.4 M NaOH. Temperature was controlled at 20 ± 1 °C. The mixed liquor was agitated at 500 rpm, except during settling and decant phases when mixing was switched off. In the aerobic phase, dissolved oxygen was controlled not to exceed 20% of saturation (around 1.8 mg/L) by an on-off valve controlling the flow of compressed air into the reactor. The HRT was controlled at 12 hours while the SRT was controlled at 8 days, not taking into account potential loss of biomass through solids in the effluent during effluent removal and biofilm removal during maintenance of the system.

5.2.1.3 Operation in anaerobic/anoxic/oxic mode

During the second operation mode, the operation of the SBR was similar to that of the first experimental phase with the difference that an anoxic phase was introduced in the parent SBR. The resulting cycle consisted of a 60 min anaerobic phase, a 105 min anoxic phase, a 105 min aerobic phase and a 90 min settling and decanting phase. To avoid oxygen intrusion, N2 was sparged for 165 min during both anaerobic and anoxic phases. In the anoxic phase, the addition of nitrate was done in a stepwise manner starting with 4 mg NO_3-N/L (final concentration) during the initial three days which was then increased to 8 mg NO_3-N/L for the remaining eleven days.

5.2.1.4 Synthetic medium

The concentrated medium was prepared in two separate solutions prepared with de-mineralized water and autoclaved at 110 °C for one hour. The first solution contained HAc and HPr as carbon source, while the other contained the required nutrients, minerals and trace elements. After dilution with water, the influent of the reactor contained per liter: 637.5 mg NaAc·$3H_2O$ (9.5 C-mmol/L, 300 mg COD/L), 6.675×10^{-2} mL $C_3H_6O_2$ (2.72 C-mmol/L, 100 mg COD/L), 107 mg NH_4Cl (2 N-mmol/L), 111.3 mg $NaH_2PO_4.H_2O$ (0.81 P-mmol/L, 25 mg PO_4^{3-}-P/L) , 90 mg $MgSO_4.7H_2O$, 14 mg $CaCl_2.2H_2O$, 36 mg KCl, 1 mg of yeast extract, and in addition, 2 mg of allyl-Nthiourea (ATU) was added to inhibit nitrification. The trace element solution was prepared as described by (Smolders et al., 1994). For addition of nitrate to the reactor in the experimental phase II. A concentrated nitrate solution of 1000 mg NO_3-N/L concentration was prepared by dissolving 7.218 g of KNO_3 in 1L of demineralized water and autoclaved at 110 °C for 30 minutes.

5.2.1.5 Monitoring of biomass performance

The performance of SBR was regularly monitored by measuring orthophosphate (PO_4^{3-}-P), acetate (Ac-C), mixed liquor suspended solids (MLSS) and mixed liquor volatile suspended solids (MLVSS), in addition to NO_3-N in second experimental phase. Pseudo steady-state condition in the SBR was confirmed based on the regular performance monitoring and daily observation of pH and DO profiles. Cycle measurements were carried out to determine the biomass activity when SBR reached pseudo steady-state conditions at the end of each experimental phase.

5.2.2 Experimental phase I: Oxidative activity withdifferent electron donors (HAc, HPr) and acceptors (O2, NO3, NO2) of sludge cultivated in anaerobic/oxic mode

Short term batch experiments were run in two double-jacketed laboratory reactors with a maximum operating volume of 0.5 L. Experiments were performed at a controlled temperature of 20±0.5 °C and automatic pH control (7.0±0.1) as both reactors were controlled by an Applikon ADI controller. Experiments were carried out either with acetate as the main carbon source or with propionate while electron acceptors varied for both cases being oxygen, nitrate and nitrite. For the batch experiments, an influent was prepared with the same composition and concentration as in the parent reactor, accept for VFA which was reduced two times to keep the same feed to biomass (VFA/TSS) ratio.

When the reactor reached steady state conditions 150 ml of sludge was transferred from the parent SBR at the end of aerobic phase, under mixing conditions, to each batch reactor. For each electron acceptor (O_2, NO_3^-, NO_2^-), two experiments were carried out in parallel, under same operational conditions except for the VFA solution which was either HAc or HPr. Each experiment lasted for 4.5 hrs, consisting of 75 min anaerobic, and 195 min either aerobic or anoxic according to the experiment purpose. In total, 18 samples were taken during each experiment. Samples were directly filtered through 0.45μL pore size filters, and kept on ice for VFA, PO_4^{+3}, NO_2^-, NO_3^- analysis.

In all tests, nitrogen gas was sparged throughout the execution of the test, except for the test with oxygen where in the aerobic phase nitrogen gas was replaced by air with a flow rate of 10 L/hour. If an anoxic phase is to follow after the anaerobic phase, either NO_2^- or NO_3^- was added. For these experiments NO_2^- and NO_3^- were introduced manually in pulses of around 10 mg NO_2-N/L and 40 mg NO_3-N/L by adding 8.5 ml of a freshly prepared nitrite and nitrate stock solutions with a concentration of 250 mg NO_2-N/L and 1000 mg NO_3-N/L respectively. To decide whether a new pulse was needed NO_2-N and NO_3-N were measured using Sigma-Aldrich nitrite test strips every 20 minutes to confirm its availability, a new injection was added in case of full consumption of electron acceptor (i.e. no color change to the strip).

5.2.3 Experimental phase II: Oxidative activity with different electron donors (HAc, HPr) and acceptors (O2, NO3, NO2) of sludge cultivated in anaerobic/anoxic/oxic mode

Two weeks after introducing an anoxic phase to the parent SBR, short-term batch experiments were carried out again following the same procedure as described in section 2.3. However, the anaerobic phase was shortened to 45 min and an extended anoxic phase of 360 min was used instead to give an opportunity to observe probable changes. Through the 405 minutes of the experiment 19 samples were collected.

5.2.4 Parameters of interest

Phosphorus was measured in all the collected samples, while the volatile fatty acids were only measured in the anaerobic phase samples. Dissolved oxygen was measured for aerobic phases and total oxygen consumption was determined by following the procedures of Smolders et al. (1994) and Brdjanovic et al. (1997). During the anoxic phases NO_2^- was always measured and NO_3^- was only measured if added.

5.2.5 Microbial community investigation

An estimation of the degree of enrichment of the bacterial populations of interest (PAO) was undertaken via fluorescence in situ hybridization (FISH) analyses. All bacteria were targeted by the EUB338mix (general bacteria probe) (Amann et al., 1990; Daims et al., 1999). 'Candidatus Accumulibacter Phosphatis' was targeted by PAOMIX probe (mixture of probes PAO462, PAO651 and PAO846) (Crocetti et al., 2000) whereas GAOMIX probe (mixture of probes GAOQ431 and GAOQ989) (Crocetti et al., 2002) was used to target 'Candidatus Competibacter Phosphatis'. For the identification of the specific PAO clades, PAO I (clade IA and other type I clades) and PAO II (clade IIA, IIC and IID) were targeted by the probes Acc-1-444 and Acc-2-444 (Flowers et al., 2009), respectively. Samples were prepared following the procedures described by (Winkler et al., 2011). Quantification of the microbial communities by FISH microscopy was conducted following the procedures described by Welles et al. (2014).

To identify the dominant microbial populations and to verify if no microbial change occurred at the subclade level, at the end of each experimental phase samples were taken and analysed with 16S-rDNA-PCR denaturing gradient gel electrophoresis (DGGE). DNA extraction, PCR amplification (Polymerase chain reaction), DGGE (Denaturing Gradient Gel Electrophoresis), band isolation, sequencing and identification of microorganisms were carried out according to the procedures described by (Bassin et al., 2011).

For thin sectioning electron microscopy, the cells were first fixed in 3% (v/v) glutaraldehyde for 1 h on ice, then postfixed in 1% (w/v) OSO_4 + 0.5 M NaCl for 3 h at room temperature, washed and stained overnight with 1% (w/v) uranyl acetate, dehydrated in ethanol series and embedded in Epoxy resin. The thin sections were finally stained with 1% lead acetate.

5.2.6 Analytical Procedure

PO_4-P was determined by the ascorbic acid method, NO_2-N and NH4-N by spectrophotometric method, and NO_3-N by the means of Dionex ICS-1000 Ion Chromatography system run by Chromeleon software. Analysis, including MLSS and MLVSS determination, were performed in accordance with Standard Methods (APHA, 1994). Dissolved oxygen concentration was automatically stored through MultiLab® Pilot software

5.3 Results

5.3.1 PAO Enrichment and Cultivation under anaerobic/oxic and anaerobic/anoxic/oxic cultivation mode

5.3.1.1 Experimental phase I - Enrichment and performance under anaerobic/oxic mode in an SBR fed with a mixture of HAc and HPr

In experimental phase I, the sequencing batch reactor for PAO enrichment was operated for more than400 days under Anerobic/Oxic operational mode fed with a mix of HAc and HPr in the ratio of 3:1. Pseudo steady-state conditions were confirmed after 60 days by the stable on-line pH and DO profiles together with stable off-line orthophosphate and MLSS measurements. Figure 5.1 shows two characteristic cycles measured on day 428 and day 477 before and after nitrate acclimatization. During the cycle on day 428 in Figure 5.1a, the sludge exhibited full COD removal in the anaerobic phase, in 20 minutes. The average orthophosphate concentration at the end of the anaerobic phase was 114.6 mg PO_4-P/L (SE = 4.43 mg PO_4-P/L, n=22) and the average P/VFA ratio was 0.61 P-mol/C-mol (SE = 0.03 P-mol/C-mol, n=22). The average orthophosphate concentration at the end of the aerobic phase was around 2 mg PO_4-P/L (SE. = 0.58 mg PO_4-P/L, n=22), leading to a stoichiometry of 17.4 mgCOD/mgP for VFA in the influent to P-removed. The MLSS and MLVSS concentrations at the end of the aerobic phase were 2.7 g/L (SE = 0.05 g/L, n=22) and 1.7 g/L (SE = 0.02 g/L, n=22), respectively. The MLVSS/MLSS ratio was 0.63 (SE = 0.009, n=22), indicating a high ash content which is often associated with a high poly-P content in enriched PAO cultures (Smolders et al., 1995).

5.3.1.2 Experimental phase II - Performance under anaerobic/anoxic/oxic cultivation mode in a SBR fed with a mixture of HAc and HPr

In experimental phase II, under Anaerobic/Anoxic/Oxic mode, the nitrate was increased in two steps, starting with 4 mg NO_3-N/L during the first three days, followed by a further increase to 8 mg NO_3-N/L, which remained for the rest of the acclimatization period. In the first three days, the 4 mg NO_3-N/L were almost all consumed in the anoxic phase, but after a further increase to 8 mg NO_3-N/L, only 4 mg NO_3-N/L were taken up for the remaining 11 days of the acclimatization period. After a total acclimatization period of 2 weeks, a cycle was closely monitored on the day 477 of operation, shown in Figure 5.1b.

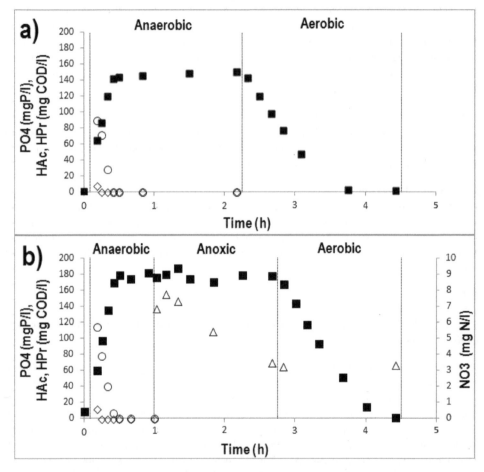

Figure 5.1 Biochemical conversions of enriched PAO I culture during characteristic cycles in phase I (on day 428) operated in anaerobic/oxic mode (a) and phase II (on day 477) operated under anaerobic/anoxic/oxic mode (b), whith PO_4 (■), HAc (○), HPr (◊),and NO_3 (△).

5.3.2 Microbial Analysis

5.3.2.1 Microbial community

The composition of the microbial community of the enriched biomass was assessed at the end of the two of the experimental phases (I and II) with both anaerobic/oxic and anaerobic/anoxic/oxic conditions. FISH analysis showed that at the end of experimental phase I, the biomass was highly enriched with PAO, while GAO were not detected (Figure 5.2a). More specifically, only PAO I was observed (Figure 5.2b), which had a coccus morphology and contained by the end of aerobic phase 1 or 2 poly-P inclusions per cell (Figure 5.2d and 5.2e). During the preparation of the thin sections, the poly-P granules disappeared from the cells, leaving big white holes in Figure 5.2d. Additional FISH quantification showed that the total bacterial population comprised of 99% PAO (SE 1.9%, n=18).

To confirm the observations of the FISH analysis and to detect any potential population change after switching from anaerobic/oxic to anaerobic/anoxic/oxic conditions, DGGE analysis was conducted (Figure 5.2c, Table 5.1). This analysis confirmed that PAO I (bands 34 and 3) was the dominant population and that PAO II and GAO were not present in both experimental phases. The DGGE analysis also showed that the side populations seen by FISH microsopy (+/-1%) and recognized in the phase contrast images by their different dimensions and morphology (Figure 5.2e), were Flavobacterium (bands 30 and 2), Chlorobi (band 31), Armatimonadetes (bands 32 and 5), Chloroflexi (bands 33, 35, 36), and Sphingomanadaceae (bands 1 and 4). No significant changes were observed in the dominant bands before and after acclimatization. However, the intensity of two minor (5 and 32) bands changed relatively, which might reflect a change in the abundance of two side populations, which may be associated with the gradual increase in the denitrification activity, observed in the beginning of the anoxic stage of each cycle, during anaerobic/anoxic/oxic operation. However FISH analysis confirmed that the system remained highly enriched by PAO I.

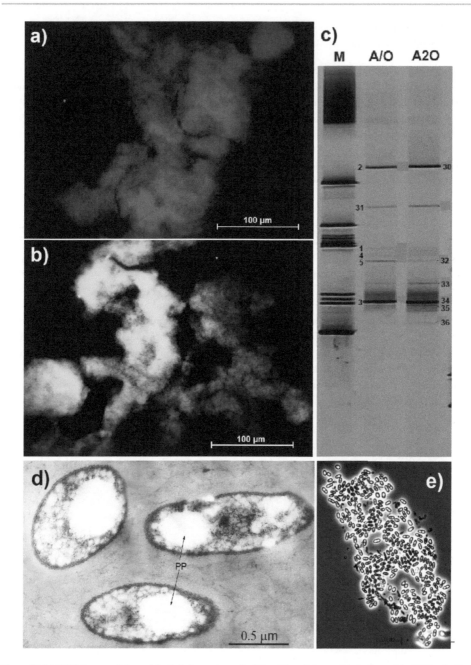

Figure 5.2 (a) FISH images of the enriched sludge at the end of Exp. Phase I (anaerobic/oxic mode): EUB mix (Cy5) cells labelled in blue, PAO mix (Cy3) labelled in red, GAO mix (Fluos) in green. (b) FISH images of the enriched sludge at the end of Exp. Phase I (anaerobic/oxic mode) PAO mix (Cy5) cells are labelled in blue, Acc II (Cy3) are in red, Acc I (Fluos) in green. (c) Image of 16S rRNA DGGE banding patterns obtained from samples at the end of Exp. Phase I (anaerobic/oxic mode) and at the end of Exp. Phase II (anaerobic/anoxic/oxic mode). Phylogenetic analysis of band sequences is shown in Table 5.1. (d) Electron microscope image of a thin section showing the poly-P (PP), and glycogen organization in the enriched PAO I cell at the end of the aerobic phase at the end of experimental phase I. (e) Phase contrast image obtained at the end of Exp. phase 1 at the end of the aerobic phase.

RDP classification						
Band	Accession No.	Phylum	Class	Order	Family	genus
1	KM884820	Ignavibacteriae	Ignavibacteria	Ignavibacteriales	Ignavibacteriaceae	Ignavibacterium
2	KM884821	Proteobacteria	Alphaproteobacteria	Sphingomonadales	Sphingomonasdaceae	Sphingopyxis
3	KM884822	Bacteroidetes	Sphingobacteriia	Sphingobacteriales	Chitinophagaceae	-
4	KM884823	Proteobacteria	Betaproteobacteria	Rhodocyclales	Rhodocyclaceae	-
5	KM884824	Armatimonadetes	-	-	-	Armatimonadetes_gp5
30	KR007577	Bacteroidetes	Sphingobacteriia	Sphingobacteriales	Chitinophagaceae	-
31	KR007578	Ignavibacteriae	Ignavibacteria	Ignavibacteriales	Ignavibacteriaceae	Ignavibacterium
32	KR007579	Armatimonadetes	-	-	-	Armatimonadetes_gp5
33	KR007580	Proteobacteria	Betaproteobacteria	Rhodocyclales	Rhodocyclaceae	
34	KR007581	Proteobacteria	Betaproteobacteria	Rhodocyclales	Rhodocyclaceae	-
35	KR007582	Proteobacteria	Betaproteobacteria	Rhodocyclales	Rhodocyclaceae	-
36	KR007583	Chloroflexi	-	-	-	-
BLAST results						
Band	Accession No.	Closest relative			ACC	% similarity
1	KM884820	Uncultured bacterium clone V201-40			HQ114055	99
2	KM884821	Uncultured bacterium clone 1A			KJ600812	99
3	KM884822	bacterium enrichment culture SBR_L_B27			KJ395389	100
4	KM884823	bacterium enrichment culture SBR_C_B04			KJ395408	100
5	KM884824	Uncultured bacterium clone Wu-C19			KJ783095	96
30	KR007577	bacterium enrichment culture SBR_L_B27			KJ395389	100
31	KR007578	Uncultured bacterium clone V201-40			HQ114055	99
32	KR007579	Bacterium enrichment culture clone SBR_S_B05			KM884824	95
33	KR007580	Uncultured bacterium clone BJ 2-72			KC551759	96
34	KR007581	bacterium enrichment culture SBR_C_B04			KJ395408	98
35	KR007582	Uncultured bacterium clone FR1-70			JX431981	92
36	KR007583	Uncultured bacterium clone SBRAC27			HQ158630	99

Table 5.1 Phylogenetic analysis of 16S rRNA sequences obtained from DGGE band profiles shown in Figure 5.2c.

5.3.3 Activity tests to assess the aerobic activity of highly enriched PAO with different e-donors

For comparison purposes with the nitrate and nitrite tests, aerobic batch tests were conducted after feeding with either HAc or HPr to determine the aerobic rates and stoichiometric values (Figure 5.3a and 5.3b). Both reactors experienced similar anaerobic VFA uptake, aerobic P-uptake, max specific oxygen uptake rates, however, under anaerobic conditions higher P-release was observed in the acetate reactor as shown in Table 5.2.

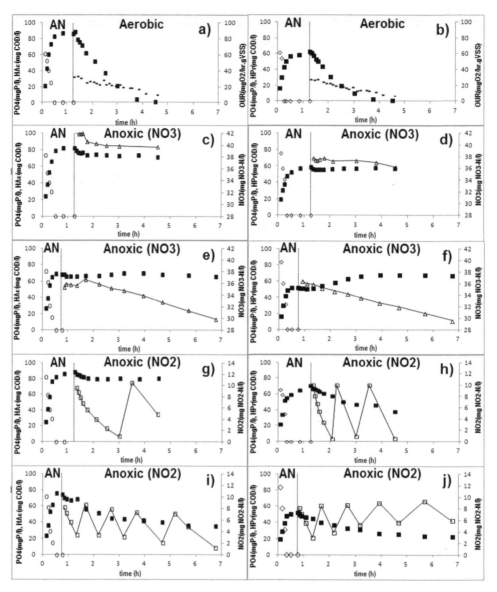

Figure 5.3. Batch test profiles showing biochemical conversions during the anaerobic (fed with either HAc or HPr), anoxic (supplied with either NO_2^- or NO_3^-) or oxic stages of PAO I enrichment culture, withdrawn from the parent reactor during operation in Anaerobic/Oxic mode (Phase I) or Anaerobic/Anoxic/Oxic mode (Phase II). (a) Anaerobic (HAc)/Aerobic batch test with Anaerobic/Oxic biomass, (b) Anaerobic (HPr)/Aerobic batch test with Anaerobic/Oxic biomass, (c) Anaerobic (HAc)/Anoxic (NO_3^-) batch test with Anaerobic/Oxic biomass, (d) Anaerobic (HPr)/ Anoxic (NO_3^-) batch test with Anaerobic/Oxic biomass, (e) Anaerobic (HAc)/ Anoxic (NO_3^-) with Anaerobic/Anoxic/Oxic biomass, (f) Anaerobic (HPr)/ Anoxic (NO_3^-) batch test with Anaerobic/Anoxic/Oxic biomass, (g) Anaerobic (HAc)/ Anoxic (NO_2^-) batch test with Anaerobic/Oxic biomass, (h) Anaerobic (HPr)/Anoxic (NO_2^-) with Anaerobic/Oxic biomass, (i) Anaerobic (HAc)/Anoxic (NO_2^-) with Anaerobic/Anoxic/Oxic biomass, (j) Anaerobic (HPr)/Anoxic (NO_2^-) with Anaerobic/Anoxic/Oxic biomass, whith: HAc (○), HPr (◊), OUR (—), NO_3^- (△), NO_2^- (□), and PO_4^{+3} (■). AN = anaerobic.

5.3.4 Activity tests to assess the denitrification activity of highly enriched PAO with different e-donors

5.3.4.1 Anoxic activity with NO₃ as an e-acceptor with biomass cultivated in anaerobic/oxic and anaerobic/anoxic/oxic mode

Batch experiments were performed at the end of experimental phase I (before acclimatization) and experimental phase II (after acclimatization), using either HAc or HPr as carbon source in each test. The biomass was fed either with HAc or HPr and subsequently exposed to NO_3^-. Almost no P-uptake was observed in all batch tests with nitrate (Figure 5.3c, d, e, f and Table 5.2). Remarkably, in the case of the propionate fed reactor in experimental phase II, after 75 mins (beginning of the anoxic stage), some P release was observed instead of uptake. Although the nitrate uptake rate increased two times during acclimatization (in experimental phase II), the anoxic respiration rate remained very low when compared to aerobic respiration as shown in Table 5.2.

5.3.4.2 Anoxic activity with NO₂ as an e-acceptor of biomass cultivated in anaerobic/oxic and anaerobic/anoxic/oxic mode

Batch experiments were performed at the end of experimental phase I (before acclimatization) and experimental phase II (after acclimatization), using either HAc or HPr as carbon source in each test. For propionate fed reactors, no change in either NO_2^- uptake or P-uptake rate occurred before and after acclimatization, indicating that the acclimatization was not needed for denitrification of NO_2^- and that the acclimatization period (with NO_3^- and presumably low concentrations of NO_2^-) had little effect on the activity. The conversions during these batch tests are shown in Figure 5.3g, h, i, j and Table 5.2. However, the acetate fed experiments showed before acclimatization no P-uptake and only half the rate of NO_2^--uptake (Figure 5.3g and h) compared to after acclimatization (Figure 5.3i and j), suggesting that the type of PHA stored in the cells, which depends on the type of carbon source fed to the system, affects the anoxic activity of the biomass or that a certain small side population with denitrifying capabilities can store PHA from propionate. Comparing the activities after acclimatization in both the acetate and propionate fed reactors, the P-uptake rate and P/N ratio in the acetate fed reactor where higher than those in the propionate fed reactor by 40% and 50%, respectively (Table 5.2).

Table 5.2 Kinetic rates and stoichiometric ratios for both HAc and HPr fed batch reactors under anaerobic, anoxic and oxic conditions with sludge cultivated in anaerobic/oxic and anaerobic/anoxic/oxic mode.

Experiments	Parameter	Biomass cultivation (Carbon source)	anaerobic/oxic (O2) anaerobic/oxic		anaerobic/anoxic (NO3) anaerobic/oxic		anaerobic/anoxic/oxic		anaerobic/anoxic (NO2) anaerobic/oxic		anaerobic/anoxic/oxic	
					HAc	HPr	HAc	HPr	HAc	HPr	HAc	HPr
Anaerobic	Max specific P release rate	mmolP/gVSS.hr	8.9	6.1	6.6	5.1	7.8	6.1	7.2	5.4	8	5.4
	Max specific VFA uptake rate	mgCOD/gVSS.hr	190	196	206	228	238	306	202	217		
	P/VFA (net P release/VFA uptake)	mmolP/mmolC	0.78	0.47	0.80	0.44	0.7	0.40	0.76	0.52	0.65	0.63
	MLVSS/MLSS end		0.72	0.70	0.69	0.69	0.69	0.68	0.71	0.70	0.55	0.4
Anoxic or, Aerobic	Max specific P uptake rate	mmolP/gVSS.hr	1.40	1.35	0	0	0	-0.21	0	0.4	0.8	0.85
	Max specific NO3 uptake rate	mmol NOx/gVSS.hr	NA	NA	0.06	0.04	0.09	0.1	0.43	0.8	NA	NA
	Max specific oxygen uptake rate	mmolO2/gVSS.hr	1	0.9	NA	NA	NA	NA	NA	NA	NA	NA
	Specific total oxygen uptake	mmolO2/gVSS	2.1	1.7	NA	NA	NA	NA	NA	NA	NA	NA
	P/N (P uptake/N uptake)	mmolP/mmolN	NA	NA	0	0	0	-2.1	0	0.51	0.70	0.46
	P/O (P uptake/Oxygen uptake)	mmolP/mmolO2	1.38	1.28	NA	NA	NA	NA	NA	NA	NA	NA
	MLVSS/MLSS		0.59	0.59	0.68	0.68	0.70	0.64	0.72	0.66	0.62	0.60

5.4 Discussion

5.4.1 Microbial community structure of biomass cultivated in anaerobic/oxic and anaerobic/anoxic/oxic mode

A PAO I culture was successfully enriched to 99% $\pm$ 1.9%, at pH 7.6 with a VFA mix of HAc and HPr 3:1 as carbon source and high influent P/C ratio (P-mol/C-mol), while no PAO II or GAO were observed. The observed P-release/VFA uptake at steady state conditions ranged between 0.62 and 0.76, indicating a high PAO enrichment with high poly-P content, which was also confirmed by the ash content of the biomass (Schuler and Jenkins, 2003; Welles et al., 2015).

Interestingly, the DGGE profiles showed that several side populations (minor bands) were present, that were also observed on the phase contrast images. Quantification of the side populations by FISH analysis indicated that those side populations formed roughly 1 % of the population. The quantification is based on relative surface areas of bacterial flocs that stain positive with certain FISH probes. As the side populations were homogenously enmeshed in the PAO dominated flocs and no specific probes were used to target those side populations, there quantity (difference between surface area of all bacteria and specific PAO/GAO) may have been underestimated. From the cycle test it was clear that all VFA was taken up during the anaerobic phase, suggesting that all carbon source was stored by PAO. The only carbon sources deliberately added to the system that may not have been taken up in the anaerobic phase by PAO is yeast extract and EDTA (added to the influent at only 1mg/L and 3mg/L respectively, together about 6 mgCOD/L). Although the acclimatization period was designed short enough to avoid major changes in the dominant microbial population, DGGE profiles indicated that a slight change in the abundance of two minor populations occurred when the systems changed from experimental phase I to experimental phase II. One band, corresponding to Chloroflexi increased slightly in abundance while another one decreased, suggesting that the Chloroflexi may be favored by the implementation of an anoxic zone. However, in past studies with Chloroflexi, its capability to denitrify has not been observed (Morgan-Sagastume et al., 2008) but known to be abundant and account for a large fraction of filaments in biological nutrient removal systems (Bjornsson et al., 2002; Morgan-Sagastume et al., 2008).

5.4.2 Aerobic activity with different electron donors

In the anaerobic phase, the observed P/VFA with HPr was lower than HAc, which is in line with previous studies (Smolders et al., 1994; Oehmen et al., 2005b; Li et al., 2008; Lopez-Vazquez et al., 2009). The sludge assimilated both acetate and propionate at similar rates as shown in Table 5.2, which agreed well with literature (Pijuan et al., 2004; Oehmen et al., 2005a). However, the P release rate was much higher in the case of HAc, and this is due to the fact that the uptake of propionate and subsequent conversion to PHA requires less amounts of energy compared to acetate (Smolders et al., 1994; Kuba et al., 1996; Oehmen et al., 2006). Regarding the difference in the intracellular PHA composition being mostly PHB in case of acetate and PHV and PH2MV in case of propionate (Oehmen et al., 2005a), PAO I can use both efficiently. It is worth to mention that from our cycle tests (measurements through one complete cycle), it is clear that all the PHA produced in one cycle through its anaerobic phase is completely exhausted by the end of the aerobic phase (data not shown) of the same cycle, which means that the PHA through the batch experiments was fully produced with either HAc or HPr in the anaerobic phase of the same experiment.

The aerobic phase did not seem to be affected by the carbon source supplied in the anaerobic phase. Both systems took up phosphorus and oxygen at similar rates. In general, the sludge performance agreed well with other studies, which confirmed that the sludge was normal and further conclusions can be built up, based on observations.

5.4.3 Denitrification activity with NO$_3$ and different electron donors

In the batch experiments with nitrate, either due to the observed change in side population or the increase in its activity after acclimatization, the nitrate uptake rates slightly increased. However, there was no significant P-uptake in all cases, suggesting there was no denitrification by PAO's. Moreover, in the case with propionate after acclimatization even a P release took place which is similar to previous observations (Kuba et al., 1993; Ahn et al., 2001; Lanham et al., 2011). In the studies of (Ahn et al., 2001), it was suggested that these observations were due to pH increase caused by denitrification, but since pH was controlled in this study, it can be concluded that the pH did not play a role. Since propionate was depleted it could also not be due to anoxic substrate uptake by PAO (Kuba et al., 1993). Overall the low denitrification activity and minimal P-uptake or P-release, suggest that PAO I cannot denitrify nitrate.

5.4.4 Is PAO I able to denitrify using nitrate?

As several studies have shown that biomass enriched under anaerobic/oxic conditions, only needed few hours (5 hrs) of acclimatization to nitrate in order to obtain anoxic denitrification and P-uptake activity (Zeng et al., 2003b), the acclimatization period of two weeks applied in this study should have been sufficient. The very low denitrification activity of (0.1 mmol NO$_3$-N/L.hr) and absence of phosphate uptake in the presence of nitrate, in comparison to the significantly higher anoxic denitrification and P-removal activity in the presence of nitrite (0.8-0.85 mmol NO$_2$-N/L.hr, 0.4-0.55 mmol P/L.hr) and aerobic oxygen consumption and P-removal activity (0.9-1 mmol O/L.hr equivalent to 0.7-0.8 mmol NO$_3$-N/L.hr, 1.35-1.4 mmol P/L.hr), as well as the high denitrification activity observed in the studies of (Kuba et al., 1993) (0.6 mmol NO$_3$-N/L.hr, 0.54 mmol P/L.hr) suggests that the PAO I, enriched in this study are not able to denitrify via NO$_3$- or that they have a very suboptimal respiratory chain for NO$_3$- reduction. Considering that denitrifying ordinary heterothrophic organisms have denitrification rates ranging from 0.14 to 2.14 mmol NO$_3$-N/(gVSS.h), depending on their carbon source (Henze et al., 2008), a 4% side population of with denitrifying ordinary heterothrophic organisms could potentially be responsible for the observed denitrification rates on nitrate in this study.

A few other studies found through a variety of different type of experiments that PAO are not able to take up nitrate (Wentzel et al., 1988; Zeng et al., 2003a; Martin et al., 2006; Guisasola et al., 2009; Kim et al., 2013). It was found that both PAOI and II can only denitrify from nitrite and never from nitrate where clades IA, IIF and IIC were tested with FISH/microautoradiography (MAR) analysis (Kim et al., 2013). Clades IA, IB, IIA (Mao et al., 2014) have been indicated to lack nitrate reduction capability through an integrated metatranscriptomic and metagenomic analysis. The only study conducted with a highly enriched PAO I culture (90%) to test denitrification over nitrate (Lanham et al., 2011), has unfortunately not reported kinetic rates of nitrate or phosphate uptake, and neither stoichiometry, besides the rather long anoxic phase of around 30 hours, without any means reported to prevent oxygen intrusion (nor redox potential measurements), implies that oxygen intrusion might have occurred causing the very slow P-uptake they have observed through the phase which was meant to be anoxic.

Several metagenomic studies generated '*Candidatus* Accumulibacter phosphatis' metagenomes from 5 different PAO clades (IA, IB, IC, IIA, IIC, IIF) (Martin et al., 2006; Flowers et al., 2013; Mao et al., 2014; Skennerton et al., 2015). These studies revealed that only the metagenome of PAO clade IIF encodes the nitrate reductase and nitrite reductase genes, whereas the metagenomes of the other clades (IA, IB, IC, IIA, IIF) only encoded genes for periplasmic nitrate reductase and nitrite reductase. The functionality of periplasmic nitrate reductase in anoxic respiration is unclear. Periplasmic nitrate reductases play different physiological roles in different type of organisms and it was observed in previous studies that the activity of periplasmic nitrate reductase was not sufficient to support anoxic respiration in several microorganisms (Moreno-Vivián et al., 1999; Skennerton et al., 2015).

The very slow denitrification activity observed in this study may be due to slow intrinsic PAO denitrification activity with a suboptimal respiratory chain or, in our opinion more likely, due to the activity of side populations.

5.4.5 Simultaneous NO₃ reduction and P-uptake - the role of side populations

Considering the findings in this study and some previous studies, suggesting that PAO I are not capable of effectively using nitrate, and the findings in some other studies that indicate that PAO II are not capable of using nitrate, it is remarkable that P-uptake was still observed in other studies in the presence of nitrate in systems dominated sometimes by PAO I (Carvalho et al., 2007; Flowers et al., 2009; Oehmen et al., 2010a; Lanham et al., 2011), and other times by PAO II as well (Guerrero et al., 2012; Wang et al., 2014). Besides, a significant activity was observed in studies that had no microbial characterization performed (Kuba et al., 1993; Ahn et al., 2001; Hu et al., 2002; Freitas et al., 2005; Parco et al., 2007) (Table 5.3). One possible explanation for these contradictions could be that in all past studies, side populations are responsible for a partial denitrification from nitrate to nitrite. This would imply that the PAO specific PO_4 uptake activity might be dependent on the NO_2^- production by another group of bacteria. This NO_2^- formation may be dependent on the abundance and type of side populations as well as other external factors (such as the availability and type of BOD) that determine the performance of those side populations.

In several long-term studies, a drop in the percentage of PAO in the culture was observed in several long-term studies (Carvalho et al., 2007; Flowers et al., 2009; Guisasola et al., 2009), associated with improved denitrifying EBPR activity, when NO_3^- was introduced, suggesting that nitrate addition supported the proliferation of flanking bacteria and thereby probably establishing a syntrophic relation between PAO and flanking bacteria, leading to improved phosphate uptake over nitrate reduction.

5.4.6 Where does the carbon source for flanking populations come from?

If side populations were responsible for the conversion of NO_3^- to NO_2^-, it remains a question, what was the source of BOD for those organisms if the VFA fed to most EBPR systems were fully taken up during the anaerobic phase. However, the fact that side populations exist in the sludge in many past studies implies that, they should have access to a carbon source otherwise they could not compete and survive. One possible carbons source could be PHA stored by denitrifying GAO and another possibility could be BOD released as soluble microbial products or from decaying biomass.

5.4.6.1 PHA storage by GAO

In studies where GAO were present, potential denitrifying GAO may have stored VFA in the form of PHA, which could have served as source of BOD during the denitrification. Several studies reported low percentage of enrichment, where GAO or many other organisms were available (Guisasola et al., 2009; Vargas et al., 2011; Winkler et al., 2011; Wang et al., 2014). Some studies on denitrifying PAO had no microbial analysis as FISH or DGGE (Kuba et al., 1993; Ahn et al., 2001; Freitas et al., 2005). Often the observed stoichiometry and rates were used as indicators to demonstrate whether the biomass was highly enriched with PAO. Recent studies have shown however that the presence of GAOs or any other organisms that can store PHA cannot be ruled out based on the stoichiometry and rates, as anaerobic stoichiometry (P/VFA) of highly enriched PAO cultures can vary a lot from 0 to 0.8 and the ISS/TSS ratio can vary roughly from 0.03 to 60% , dependent on the types of PAO clades, P content of the cells, and pH (Welles et al., 2015). In the studies by (Kerrn-Jespersen and Henze, 1993; Kuba et al., 1993) in anaerobic/anoxic configurations with nitrate, very significant anoxic PAO activity was observed, but the anaerobic P/VFA ratios in those studies indicate that the potential presence of a 30 to 40% (enough for the partial denitrification from nitrate to nitrite) GAO fraction cannot be ruled out.

Table 5.3 Denitrifying PAO activity with different electron donors and acceptors in previous studies.

	Anoxic						Sludge characteristics							Parameter
P/N	NOx uptake rate	P uptake rate	P/VFA	VFA uptake rate	P release rate	pH	GAO	PAO II	PAO I	PAO mix	e- acceptor	e- donor	Units
mmolP/mmolN	mmolNOx-N/gVSS.hr	mmolP/gVSS.hr	mmolP/mmolC	mmolC/gVSS.hr	mmolP/gVSS.hr		%	%	%	%			
0.61	0.49	0.3	0.32			7.5	19	38	28	66	NO$_2$	HPr	(Wang et al., 2014)
0.55	0.25	0.14	0.36			7.5	27	54	21	75	NO$_3$	HPr	
0.89-0.33**	-	-	0.49	2.75	1.35	7-8.2	0	0	90	90	NO$_2$	HPr	(Lanham et al., 2011)
1-0.33**	0.23*	0.05*									NO$_3$	HPr	
0.48	1	0.48	0.55	-	-	7.5	1	-	-	40	NO$_2$	HAc	(Vargas et al., 2011)
0.34	1.2	0.41	0.38	-	-	7.5	1	-	-	60	NO$_2$	HPr	
0.63	0.154	0.067	-	-	-	7.3	-	5	67	72	NO$_3$	HAc	(Flowers et al., 2009)
0.29	0.16	0.03	-	-	-	7.3	-	50	32	82	NO$_3$	HAc	
0.3	0.69	0.21	-	-	-	7.5	3	mostly	-	30	NO$_2$	HPr	(Guisasola et al., 2009)
0.55	0.12	0.07	-	-	-						NO$_3$	HPr	
0.6	0.45	0.27	0.16			7-8.2	-	37	0.3	38	NO$_3$	HAc	(Carvalho et al., 2007, Oehmen et al., 2010a)
0.82	0.77	0.63	0.32				-	31	44	76	NO$_3$	HPr	
0.45	:-	-	-					66	34	51-72	NO$_2$	HAc	(Jiang et al., 2006)
0.51	-	-	-								NO$_3$	HAc	
0.83	0.47	0.39	0.31	1.94	0.59	7-7.5	-	-	-	-	NO$_3$	HAc	(Freitas et al., 2005)
0.5*	1.1*	0.55	0.35	1.87	0.65	7	-	-	-	38	NO$_3$	HAc	(Zeng et al., 2003b)
0.18	-	-	-	-	-	7	-	-	-	-	NO$_2$	HAc	(Ahn et al., 2001)
1	-	-	-	-	-		-	-	-	-	NO$_3$	HAc	
0.9	0.6	0.54	0.48	1.13	0.54	7-7.3	-	-	-	-	NO$_3$	HAc	(T. Kuba et al., 1993)

*Estimated from the original data, ** Corresponding to pH, - Not reported.

SMP can be produced from normal bacterial growth and metabolism (Herbert, 1961), decay and lysis (Noguera et al., 1994), or in this case, can be due to the very low concentration of NO_2^- where SMP might have been produced by some bacteria to scavenge this required and scarce nutrient (Postgate and Hunter, 1964). Heterotrophic growth on SMP was discussed in literature (Furumai and Rittmann 1992; Rittman et al., 1994). Based on the studies of (Gujer et al., 1995; Lopez et al., 2006; Hao et al., 2010; Vargas et al., 2013), the potential growth associated SMP production per cycle was estimated to be 18 mgCOD/L/cycle. Besides some other indirect carbon sources that can contribute a low COD concentration as aeration with air contaminated with COD traces (Moussa et al., 2005) or by means of media preparations in the used water, nutrient compounds and yeast extract (Rittman et al., 1994; Moussa et al., 2005).

Considering similar biomass yields for PAO and ordinary heterotrophs, the estimated COD from released SMP's, yeast extract and EDTA in the influent (together good for 18 mgCOD/L/cycle + 3mgCOD/L/cycle = 21 mgCOD/L/cycle, compared to 200 mgCOD/L/cycle), would potentially allow the proliferation of a side population of about 10% from the total bacterial population. Considering the total nitrate removal observed in this study in one cycle (4 mg NO_3-N/L) a total COD of 34.4 mg COD/L (= 8.6 mgCOD/mg NO_3-N (assuming a biomass yield of 0.67 mgVSS/mg COD) x 4 mg NO_3-N/L (Henze, 2008)) would be needed to denitrify when the full denitrification is conducted by side population and only 13.8 mgCOD/L is needed (40%) to convert nitrate to nitrite when side populations perform a partial denitrification from nitrate to nitrite, which is further converted to nitrogen gas by PAO. This suggests that side populations may play a crucial role in denitrification activity observed in this an many past studies in anaerobic/oxic systems with low N-removal rates in Anaerobic/Anoxic/Oxic systems. Considering that this nitrate was converted to nitrite by side populations (0.1 mmol NO_3-N/gVSS.L) and that nitrite was further denitrified by PAO, the supply rate of nitrite in the nitrate test was even lower than the nitrite conversion rate by PAO (0.85 mmol NO_2-N/gVSS.L) and therefore the energy production rate of PAO was probably not high enough to take up phosphate in the nitrate tests, and that's why almost no P-uptake was observed in parallel to nitrate uptake through the acclimatization period and in the batch experiments.

5.4.7 Quantitative comparison of rates and stoichiometry in literature

Considering the possibility that side populations are responsible for the partial denitrification, higher fractions of side populations in biomass microbial communities, would lead to better partial denitrification and its associated nitrite production and consequently better anoxic P-removal by PAO, but when the fractions of side populations become very big and PAO fractions become very low, the anoxic biomass P-removal activity would tend to go down due to the low PAO presence. In recent studies, FISH and or DGGE analysis were conducted, often showing that the PAO enriched microbial communities which performed denitrification with nitrate, were still containing significant fractions of organisms other than PAO. Figure 5.4a is showing the anoxic P-activity against PAO fraction observed in different studies with nitrite and nitrate, a bell-shape relationship between the PAO percentage and the anoxic P-uptake rate could be observed in the literature when system was fed with nitrate Table 5.3. The highest anoxic P-uptake activity can be observed when the PAO percentage is around half the population, while highly enriched PAO cultures show the lowest anoxic P-uptake rates.

Considering the stoichiometry, Figure 5.4b shows the relation between P/N and PAO% shows an almost flat profile at around 0.5, which means that the stoichiometry did not change much in correspondence to the fraction of PAO in the community Table 5.3. This could be due to a syntrophic relation, where PAO depend on NO_2^- production from NO_3^- reduction by denitrifying ordinary heterotrophs or denitrifying GAO.

On the contrary, Figure 5.4c displaying the P/N against PAO% with nitrite, shows an increase in stoichiometry with the increase of PAO% up to 0.83, which can be explained by the scenario that both PAO and denitrifying ordinary heterotrophs are competing for NO_2^-, so the higher the percentage of PAO in the population, the higher the observed P/N ratio. When comparing the maximum stoichiometry with nitrate and nitrite, a higher P/N stoichiometry with NO_3^- would have been expected with nitrate if PAO is making use of both NO_3^- or NO_2^-, as nitrite has a similar reduction potential, but 1 mol of NO_3^- accepts 5 e^- compared to NO_2^- that accepts only 3 e^-. Therefore, the energy produced per NO_x-N reduction would be about 5/3 times higher for nitrate allowing more P-uptake. However, the reported values of the studies shown in Figure 5.4b and c (with microbial characterization and quantification) are on the contrary (P/N around 0.5 with NO_3^-, and up to 0.83 with NO_2^-), which supports that PAO might not be able to make direct use of NO_3^-, so they might be relying on flanking organisms for NO_3^- reduction to NO_2^-.

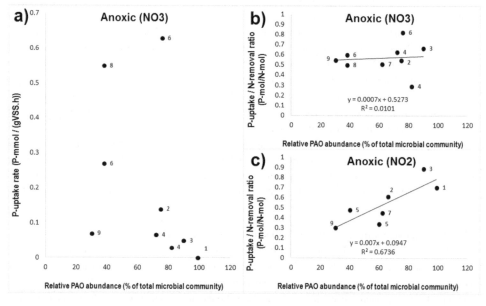

Figure 5.4 P-uptake rates and stoichiometry at different PAO enrichment percentages fed with either nitrate (a,b) or nitrite (c) in the literature. 1) This study, 2) (Wang et al., 2014), 3) (Lanham et al., 2011), 4) (Flowers et al., 2009), 5) (Vargas et al., 2011), 6) (Carvalho et al., 2007), 7) (Jiang et al., 2006) 8) (Zeng et al., 2003b), 9) (Guisasola et al., 2009). Only those studies with a community analysis have been considered.

This indicates that in the studies with relatively low denitrification activity, the nitrate might not be taken up by PAO I and neither PAO II, but by flanking bacteria which live on the COD brought from the decay of dead cells, or by denitrifying GAOs (Third et al., 2003; Zeng et al., 2003a; Zeng et al., 2003c; Martin et al., 2006; Guisasola et al., 2009; Oehmen et al., 2010b; Winkler et al., 2011; Kim, et al., 2013) or some other denitrifying heterotrophic bacteria able to store substrate anaerobically. It also helps to explain why some cultures show anoxic P-uptake and others not independent of the type of PAO clade. In general, studies that applied molecular tools in systems running under altering anaerobic and aerobic conditions performing simultaneous N and P removal are more concerned about the main PAO population changes, with less attention to the side population changes (Ahn et al., 2001; Kong et al., 2002; Zeng et al., 2003b).

5.4.8 Denitrification activity with NO_2 as an e-acceptor with different C-sources

Acclimatization seemed to be not essential for both the HAc and HPr fed reactors, as denitrification activity was observed in both reactors although a faster rate was observed in the HPr fed experiment. Before acclimatization, in the HAc fed reactor there was low nitrite removal rate for unknown reasons and no P-uptake while in the HPr fed reactor there was a faster nitrite removal rate with P-uptake, which has also been observed in other studies (Guisasola et al., 2009; Vargas et al., 2011; Tayà et al., 2013; Wang et al., 2014). Possibly the low nitrite removal rate in HAc reactor did not result in energy production rate that could support metabolic processes other than maintenance (Carvalho et al., 2007; Vargas et al., 2011).

After acclimatization, the HAc and HPr fed reactors had similar denitrification rates but the P-uptake rate, as well as the P/N ratio were higher in the HAc fed reactor. This can be explained as follows: in these experiments, the biomass from the acetate fed reactor used more poly-P per carbon VFA and also more glycogen per carbon VFA in the anaerobic phase in comparison to the HPr fed reactor, moreover the ratio of poly-P consumption to glycogen consumption was higher in the HAc uptake (Oehmen et al., 2005a). Therefore, the HAc fed reactor may channel a bigger fraction of its energy generated by denitrification to P-uptake and possibly to glycogen uptake, while the HPr fed reactor may use most of the energy for growth.

5.4.9 Implications/recommendations

The failure or the fluctuation in the performance of real wastewater treatment plants in performing simultaneous biological nitrogen and phosphorus removal can be resolved by a better understanding of the PAO denitrification capacities. The results in the literature are not unequivocal that the PAO can directly use nitrate, and instead it only uses nitrite reduced from other microorganisms. When either the external carbon source for denitrifying ordinary heterotrophic organisms or the internal carbon source for non-PAO carbon storing denitrifying organisms (like GAO) is depleted both nitrate and P consumptions may be hindered even if the PHA levels in PAO are sufficient. This implies that in real WWTP, PAO may experience anoxic P release if the nitrite concentrations became too low, either due to lack of COD available for denitrifying ordinary heterotrophs or their inhibition by toxic compounds, or on the other hand if VFA enters the anoxic stage.

More investigations should be done to assess the role of side population in denitrification and to identify the microbial structure of the flanking bacteria and their symbiosis with PAO population, the mechanism of electron transfer from the electron donor to acceptor in those microorganisms. Adding a carbon source that is not used by neither PAOs nor GAOs can trigger the proliferation of side population to a pronounced percentage to facilitate the study of their structure and contribution to denitrification.

The role of DGAO can neither be confirmed nor ignored according to this study or some similar studies having no GAO population (Carvalho et al., 2007; Oehmen et al., 2010a; Lanham et al., 2011; Vargas et al., 2011; Guerrero et al., 2012), consequently we recommend a deeper investigation in the role of DGAO, and their ability of denitrification of nitrate, specially that GAO are considered to be the natural competitor to PAO and they are abundant in all phosphate removal systems competing for VFAs without contribution to phosphorus removal, however, if they can indirectly contribute to phosphorus removal through nitrate reduction, it will be beneficial to have GAO in the system (Winkler et al., 2011).

The symbiosis with other microorganisms should be further studied focusing on the rates of denitrification from nitrate to nitrite by flanking bacteria, and from nitrite to nitrogen by either PAO or flanking bacteria along with modelling the competition between them for nitrite.

The final product of denitrification can be either nitrogen or nitrous oxide (Wang et al., 2014). Nitrous oxide can be more toxic to PAO II compared to PAO I causing this claim of disability of PAO II to denitrify from nitrate, which needs further investigation.

5.5 Conclusions

A highly enriched anaerobic/oxic PAO I culture was obtained on HAc/HPr as carbon substrate. The enriched anaerobic/oxic PAO I culture was not able to take up phosphate significantly in the presence of nitrate, while denitrification coupled to phosphate uptake was observed in the presence of nitrite. A literature study covering studies with anaerobic/anoxic/oxic denitrifying EBPR cultures, in which the degree of PAO enrichment was assessed by FISH, demonstrated that the anoxic P-uptake rates in past studies tend to be low or almost negligible with highly enriched PAO cultures, whereas significant better P-removal activities were observed in EBPR cultures with significant fractions of bacterial populations other than PAO. In additions, it was observed in many studies that the P/N stoichiometry was lower in the presence of nitrate when compared to nitrite, supporting the hypothesis that the conversion of nitrate to nitrite and part of the nitrite to molecular nitrogen gas was converted by micro-organisms other than PAO. Based on our data and literature data the authors suggest that P-removal under anoxic conditions with nitrate might be dependent on the formation of nitrite produced by other denitrifying organisms.

5.7 References

Ahn J, Daidou T, Tsuneda S, Hirata A (2001) Metabolic behavior of denitrifying phosphate-accumulating organisms under nitrate and nitrite electron acceptor conditions. Journal of Bioscience and Bioengineering 92: 442-446

Ahn J, Daidou T, Tsuneda S, Hirata A (2002) Characterization of denitrifying phosphate-accumulating organisms cultivated under different electron acceptor conditions using polymerase chain reaction-denaturing gradient gel electrophoresis assay. Water Research 36: 403-412

Amann R, Binder BJ, Olson RJ, Chisholm SW, Devereux R, Stahl DA (1990) Combination of 16S rRNA-targeted oligonucleotide probes with flow cytometry for analyzing mixed microbial populations. Applied and environmental microbiology 56: 1919-1925

APHA (1994) Standard methods for the examination of water and wastewater : 18th edition supplement APHA-AWWA-WEF, Washington, D.C.

Bassin JP, Pronk M, Muyzer G, Kleerebezem R, Dezotti M, van Loosdrecht MC (2011) Effect of elevated salt concentrations on the aerobic granular sludge process: linking microbial activity with microbial community structure. Applied and environmental microbiology 77: 7942-7953

Bjornsson L, Hugenholtz P, Tyson GW, Blackall LL (2002) Filamentous Chloroflexi (green non-sulfur bacteria) are abundant in wastewater treatment processes with biological nutrient removal. Microbiology 148: 2309-2318

Bortone G, Saltarelli R, Alonso V, Sorm R, Wanner J, Tilche A (1996) Biological anoxic phosphorus removal - the dephanox process. Water Science and Technology 34: 119-128

Brdjanovic D, Van Loosdrecht MCM, Hooijmans CM, Alaerts GJ, Heijnen JJ (1997) Temperature Effects on Physiology of Biological Phosphorus Removal. Journal of Environmental Engineering 123: 144-153

Broughton A, Pratt S, Shilton A (2008) Enhanced biological phosphorus removal for high-strength wastewater with a low rbCOD:P ratio. Bioresource Technology 99: 1236-1241

Carvalho G, Lemos PC, Oehmen A, Reis MAM (2007) Denitrifying phosphorus removal: Linking the process performance with the microbial community structure. Water Research 41: 4383-4396

Crocetti G, Hugenholtz P, Bond PL, Schuler A, Keller J, Jenkins D, Blackall LL (2000) Identification of polyphosphate-accumulating organisms and design of 16S rRNA-directed probes for their detection and quantitation. Applied and environmental microbiology 66: 1175-1182

Crocetti GR, Banfield JF, Keller J, Bond PL, Blackall LL (2002) Glycogen-accumulating organisms in laboratory-scale and full-scale wastewater treatment processes. Microbiology 148: 3353-3364

Daims H, Brühl A, Amann R, Schleifer KH, Wagner M (1999) The domain-specific probe EUB338 is insufficient for the detection of all Bacteria: development and evaluation of a more comprehensive probe set. Systematic and applied microbiology 22: 434-444

Flowers JJ, He S, Malfatti S, del Rio TG, Tringe SG, Hugenholtz P, McMahon KD (2013) Comparative genomics of two 'Candidatus Accumulibacter' clades performing biological phosphorus removal. Isme j 7: 2301-2314

Flowers JJ, He S, Yilmaz S, Noguera DR, McMahon KD (2009) Denitrification capabilities of two biological phosphorus removal sludges dominated by different 'Candidatus Accumulibacter' clades. Environmental Microbiology Reports 1: 583-588 DOI 10.1111/j.1758-2229.2009.00090.x

Freitas F, Temudo M, Reis MM (2005) Microbial population response to changes of the operating conditions in a dynamic nutrient-removal sequencing batch reactor. Bioprocess and Biosystems Engineering 28: 199-209

Furumai H, Rittmann BE (1992) Advanced Modeling of Mixed Populations of Heterotrophs and Nitrifiers Considering the Formation and Exchange of Soluble Microbial Products. Water Science and Technology 26: 493-502

Gonzalez-Gil G, Holliger C (2011) Dynamics of microbial community structure of and enhanced biological phosphorus removal by aerobic granules cultivated on propionate or acetate. Applied and environmental microbiology 77: 8041-8051

Guerrero J, Tayà C, Guisasola A, Baeza JA (2012) Understanding the detrimental effect of nitrate presence on EBPR systems: effect of the plant configuration. Journal of Chemical Technology & Biotechnology 87: 1508-1511

Guisasola A, Qurie M, Vargas MdM, Casas C, Baeza JA (2009) Failure of an enriched nitrite-DPAO population to use nitrate as an electron acceptor. Process Biochemistry 44: 689-695

Gujer W, Henze M, Mino T, Matsuo T, Wentzel MC, Marais GvR (1995) The Activated Sludge Model No. 2: Biological phosphorus removal. Water Science and Technology 31: 1-11

Hao X, Wang Q, Cao Y, van Loosdrecht MCM (2010) Experimental evaluation of decrease in the activities of polyphosphate/glycogen-accumulating organisms due to cell death and activity decay in activated sludge. Biotechnology and Bioengineering 106: 399-407

Henze M (2008) Biological wastewater treatment : principles, modelling and design IWA Pub., London

Henze M, Poul Hs, Jes la Cour J, Erik A (1997) Wastewater treatment : biological and chemical processes, second ed. edn Springer, Berlin; New York

Herbert H (1961) The chemical composition of microorganisms as a function of their environment. Cambridge University Press for Society for General Microbiology 11: 391-416

Hu Z, Wentzel MC, Ekama GA (2002) Anoxic growth of phosphate-accumulating organisms (PAOs) in biological nutrient removal activated sludge systems. Water Research 36: 4927-4937

Jiang Y, Wang B, Wang L, Chen J, He S (2006) Dynamic Response of Denitrifying Poly-P Accumulating Organisms Batch Culture to Increased Nitrite Concentration as Electron Acceptor. Journal of Environmental Science and Health, Part A 41: 2557-2570

Kerrn-Jespersen JP, Henze M (1993) Biological phosphorus uptake under anoxic and aerobic conditions. Water Research 27: 617-624

Kim J, Lee HJ, Lee DS, Jeon CO (2013) Characterization of the denitrification-associated phosphorus uptake properties of "Candidatus Accumulibacter phosphatis" clades in sludge subjected to enhanced biological phosphorus removal. Applied and Environmental Microbiology 79: 1969-1979

Kong Y, Ong SL, Ng WJ, Liu W-T (2002) Diversity and distribution of a deeply branched novel proteobacterial group found in anaerobic-aerobic activated sludge processes. Environmental Microbiology 4: 753-757

Kuba T, Murnleitner E, van Loosdrecht MC, Heijnen JJ (1996) A metabolic model for biological phosphorus removal by denitrifying organisms. Biotechnology and bioengineering 52: 685-695

Kuba T, Smolders G, van Loosdrecht MCM, Heijnen JJ (1993) Biological Phosphorus Removal from Wastewater by Anaerobic-Anoxic Sequencing Batch Reactor. Water Science and Technology 27: 241-252

Lanham A, Moita R, Lemos PC, Reis MA (2011) Long-term operation of a reactor enriched in Accumulibacter clade I DPAOs: performance with nitrate, nitrite and oxygen. Water science and technology 63: 352-359

Li H, Chen Y, Gu G (2008) The effect of propionic to acetic acid ratio on anaerobic–aerobic (low dissolved oxygen) biological phosphorus and nitrogen removal. Bioresource Technology 99: 4400-4407

Lopez-Vazquez CM, Oehmen A, Hooijmans CM, Brdjanovic D, Gijzen HJ, Yuan Z, van Loosdrecht MCM (2009) Modeling the PAO–GAO competition: Effects of carbon source, pH and temperature. Water Research 43: 450-462

Lopez C, Pons MN, Morgenroth E (2006) Endogenous processes during long-term starvation in activated sludge performing enhanced biological phosphorus removal. Water Research 40: 1519-1530

Mao Y, Yu K, Xia Y, Chao Y, Zhang T (2014) Genome reconstruction and gene expression of " candidatus accumulibacter phosphatis" Clade IB performing biological phosphorus removal. Environmental Science and Technology 48: 10363-10371

Martin HG, Ivanova N, Kunin V, Warnecke F, Barry KW, McHardy AC, Yeates C, He S, Salamov AA, Szeto E (2006) Metagenomic analysis of two enhanced biological phosphorus removal (EBPR) sludge communities. Nature Biotechnology 24: 1263-1269

Meinhold J, Filipe CDM, Daigger GT, Isaacs S (1999) Characterization of the denitrifying fraction of phosphate accumulating organisms in biological phosphate removal. Water Science and Technology 39: 31-42

Mino T, Arun V, Tsuzuki Y, Matsuo T (1987) Effect of phosphorus accumulation on acetate metabolism in the biological phosphorus removal process. In: Ramadori RE (ed) Biological Phosphate Removal from Wastewaters:27-38.

Mino T, van Loosdrecht MCM, Heijnen JJ (1998) Microbiology and biochemistry of the enhanced biological phosphate removal process. Water Research 32: 3193-3207

Moreno-Vivián C, Cabello P, Martínez-Luque M, Blasco R, Castillo F (1999) Prokaryotic Nitrate Reduction: Molecular Properties and Functional Distinction among Bacterial Nitrate Reductases. Journal of Bacteriology 181: 6573-6584

Morgan-Sagastume F, Nielsen JL, Nielsen PH (2008) Substrate-dependent denitrification of abundant probe-defined denitrifying bacteria in activated sludge. FEM FEMS Microbiology Ecology 66: 447-461

Moussa MS, Hooijmans CM, Lubberding HJ, Gijzen HJ, van Loosdrecht MC (2005) Modelling nitrification, heterotrophic growth and predation in activated sludge. Water research 39: 5080-5098

Noguera DR, Araki N, Rittmann BE (1994) Soluble microbial products (SMP) in anaerobic chemostats. Biotechnology and Bioengineering 44: 1040-1047

Oehmen A, Carvalho G, Freitas F, Reis MAM (2010a) Assessing the abundance and activity of denitrifying polyphosphate accumulating organisms through molecular and chemical techniques. Water Science and Technology 61: 2061-2068

Oehmen A, Carvalho G, Lopez-Vazquez CM, van Loosdrecht MCM, Reis MAM (2010b) Incorporating microbial ecology into the metabolic modelling of polyphosphate accumulating organisms and glycogen accumulating organisms. Water Research 44: 4992-5004

Oehmen A, Lemos PC, Carvalho G, Yuan Z, Keller J, Blackall LL, Reis MAM (2007) Advances in enhanced biological phosphorus removal: From micro to macro scale. Water Research 41: 2271-2300 Oehmen A, Saunders AM, Vives MT, Yuan Z, Keller J (2006) Competition between polyphosphate and glycogen accumulating organisms in enhanced biological phosphorus removal systems with acetate and propionate as carbon sources. Journal of Biotechnology 123: 22-32

Oehmen A, Yuan Z, Blackall LL, Keller J (2005a) Comparison of acetate and propionate uptake by polyphosphate accumulating organisms and glycogen accumulating organisms. Biotechnology and Bioengineering 91: 162-168

Oehmen A, Zeng RJ, Yuan Z, Keller J (2005b) Anaerobic metabolism of propionate by polyphosphate-accumulating organisms in enhanced biological phosphorus removal systems. Biotechnology and Bioengineering 91: 43-53

Parco V, du Toit G, Wentzel M, Ekama G (2007) Biological nutrient removal in membrane bioreactors: denitrification and phosphorus removal kinetics. Water Sci Technol 56: 125-134

Pijuan M, Saunders AM, Guisasola A, Baeza JA, Casas C, Blackall LL (2004) Enhanced biological phosphorus removal in a sequencing batch reactor using propionate as the sole carbon source. BIT Biotechnology and Bioengineering 85: 56-67

Postgate JR, Hunter JR (1964) Accelerated Death of Aerobacter aerogenes Starved in the Presence of Growth-Limiting Substrates. Journal of General Microbiology Journal of General Microbiology 34: 459-473

Rittman BE, Regan JM, Stahl DA (1994) Nitrification as a source of soluble organic substrate in biological treatment. Water science and technology : a journal of the International Association on Water Pollution Research 30: 1

Schuler AJ, Jenkins D (2003) Enhanced biological phosphorus removal from wastewater by biomass with different phosphorus contents, Part I: Experimental results and comparison with metabolic models. Water environment research : a research publication of the Water Environment Federation 75: 485-498

Skennerton CT, Barr JJ, Slater FR, Bond PL, Tyson GW (2015) Expanding our view of genomic diversity in Candidatus Accumulibacter clades. Environmental Microbiology 17: 1574-1585

Smolders GJ, Klop JM, van Loosdrecht MC, Heijnen JJ (1995) A metabolic model of the biological phosphorus removal process: I. Effect of the sludge retention time. Biotechnology and Bioengineering 48: 222-233

Smolders GJ, van der Meij J, van Loosdrecht MC, Heijnen JJ (1994) Model of the anaerobic metabolism of the biological phosphorus removal process: Stoichiometry and pH influence. Biotechnology and Bioengineering 43: 461-470

Tayà C, Garlapati VK, Guisasola A, Baeza JA (2013) The selective role of nitrite in the PAO/GAO competition. Chemosphere 93: 612-618

Tchobanoglous G, Burton FL, Stensel H, Metcalf D, Eddy (2008) Wastewater engineering : treatment and reuse Metcalf and Eddy Inc., McGraw-Hill, Boston, [Mass.]; London

Third KA, Burnett N, Cord-Ruwisch R (2003) Simultaneous nitrification and denitrification using stored substrate (phb) as the electron donor in an SBR. Biotechnology and Bioengineering 83: 706-720

Thomas M, Wright P, Blackall L, Urbain V, Keller J (2003) Optimisation of Noosa BNR plant to improve performance and reduce operating costs. Water Science and Technology 47: 141-148

Vargas M, Guisasola A, Artigues A, Casas C, Baeza JA (2011) Comparison of a nitrite-based anaerobic-anoxic EBPR system with propionate or acetate as electron donors. Process Biochemistry 46: 714-720

Vargas M, Yuan Z, Pijuan M (2013) Effect of long-term starvation conditions on polyphosphate- and glycogen-accumulating organisms. Bioresour Technol 127: 126-131 Wang Y, Zhou S, Ye L, Wang H, Stephenson T, Jiang X (2014) Nitrite survival and nitrous oxide production of denitrifying phosphorus removal sludges in long-term nitrite/nitrate-fed sequencing batch reactors. Water Research 67: 33-45

Welles L, Tian WD, Saad S, Abbas B, Lopez-Vazquez CM, Hooijmans CM, van Loosdrecht MCM, Brdjanovic D (2015) Accumulibacter clades Type I and II performing kinetically different glycogen-accumulating organisms metabolisms for anaerobic substrate uptake. Water Research 83: 354-366

Wentzel MC, Loewenthal RE, Ekama GA, Marais GvR (1988) Enhanced polyphosphate organism cultures in activated sludge systems - Part 1 : Enhanced culture development. Water SA 14: 81-92

Winkler MKH, Bassin JP, Kleerebezem R, de Bruin LMM, van den Brand TPH, van Loosdrecht MCM (2011) Selective sludge removal in a segregated aerobic granular biomass system as a strategy to control PAO–GAO competition at high temperatures. Water Research 45: 3291-3299

Zeng RJ, Lemaire R, Yuan Z, Keller J (2003a) Simultaneous nitrification, denitrification, and phosphorus removal in a lab-scale sequencing batch reactor. Biotechnology and Bioengineering 84: 170-178

Zeng RJ, Saunders AM, Yuan Z, Blackall LL, Keller J (2003b) Identification and comparison of aerobic and denitrifying polyphosphate-accumulating organisms. Biotechnology and Bioengineering 83: 140-148

Zeng RJ, Yuan Z, Keller J (2003c) Enrichment of denitrifying glycogen-accumulating organisms in anaerobic/anoxic activated sludge system. Biotechnology and Bioengineering 81: 397-404

6

Impact of salinity on the anaerobic metabolism of polyphosphate-accumulating organisms (PAO) and glycogen-accumulating organisms (GAO)

Abstract

The use of saline water as secondary quality water in urban environments for sanitation is a promising alternative towards mitigating fresh water scarcity. However, this alternative will increase the salinity in the wastewater generated that may affect the biological wastewater treatment processes, such as biological phosphorus removal. In addition to the production of saline wastewater by the direct use of saline water in urban environments, saline wastewater is also generated by some industries. Intrusion of saline water into the sewers is another source of salinity entering the wastewater treatment plant. In this study, the short-term effects of salinity on the anaerobic metabolism of polyphosphate-accumulating organisms (PAO) and glycogen-accumulating organisms (GAO) were investigated to assess the impact of salinity on EBPR. Hereto, PAO II and GAO cultures enriched at a relatively low salinity level (0.02%W/V) were exposed to salinity concentrations of up to 6% (as NaCl) in anaerobic batch tests. It was demonstrated that both PAO II and GAO are affected by higher salinity levels, PAO being more sensitive to the increasing salinity. The maximum acetate uptake rate of PAO II decreased by 71% when the salinity increased from 0 to 1%, while that of GAO decreased by 41% for the same salinity increase. Regarding the stoichiometry of PAO, a decrease in the P-release/HAc-uptake ratio accompanied an increase in the glycogen-consumption/HAc-uptake ratio was observed for PAO II when the salinity increased from 0 to 2% salinity, indicating a metabolic shift from a poly-P dependent to a glycogen dependent metabolism. The anaerobic maintenance requirements of PAO II and GAO increased as the salinity concentrations risen up to 4% salinity.

Adapted from:

Welles, L., Lopez-Vazquez, C. M., Hooijmans, C. M., Van Loosdrecht, M. C. M., Brdjanovic, D. (2014). Impact of salinity on the anaerobic metabolism of phosphate-accumulating organisms (PAO) and glycogen-accumulating organisms (GAO). *Applied microbiology and biotechnology*, *98*(17), 7609-7622.

6.1 Introduction

The benefits of using saline (sea and brackish) water as secondary quality water when compared to other water production and treatment applications (including a practically infinite availability of water in the oceans) make it a promising, cost-effective and environmentally-friendly alternative to alleviate fresh water stress in urban areas located in coastal zones and inland areas with brackish ground water (Tang et al., 2006; WSD, 2009; Leung et al., 2012). From the water consumption perspective, up to 30% of the fresh water use can be replaced by saline water for certain household activities such as toilet flushing. This practice leads to the generation of saline wastewater which may affect the biological processes employed at sewage treatment plants. In addition to the direct use of saline water in urban environments, saline wastewater (containing high phosphorus concentrations) is also generated by some industries such as food production and processing (dairy, fish processing, pickled vegetables and meat canning) and tanneries (Gonzalez *et al.*, 1983; Orhon *et al.*, 1999; Fahim *et al.*, 2000; Lefebvre *et al.*, 2006). Due to tightening regulations to protect the surface water bodies against eutrophication in combination with the potential use of saline water as secondary quality water and the global growing industrialization (particularly in developing and emerging economies), the removal of nitrogen (N) and phosphorus (P) from saline wastewaters begins to attract an increasing attention.

While most of the research on saline wastewater has focused so far on the efficiency of the biological removal of organic matter and, to a lesser extent, on nitrogen removal, only a few studies have focused on the enhanced biological phosphorus removal (EBPR) process. The findings of these studies are inconsistent possibly because polyphosphate-accumulating organisms (PAO) were not the prevailing microorganisms and the operating conditions were not optimal to sustain the EBPR process. For instance, in certain cases P-removal did not occur due to EBPR but through biosynthesis (Uygur and Kargi, 2004; Kargi and Uygur, 2005; Uygur 2006), while in other studies glycogen-accumulating organisms (GAO) appeared to be the prevailing microorganisms instead of PAO (Panswad and Anan (1999). Furthermore, in some studies there was a lack of the appropriate carbon source (volatile fatty acids, VFA) in the influent (Hong *et al.*, 2007; Wu *et al.*, 2008) or the EBPR removal was affected by nitrite (NO_2^--N) and nitrate (NO_3^--N) intrusion into the anaerobic phase of the system (Intrasungkha *et al.*, 1999). Moreover, in most of the studies, information needed to assess the impact of salinity on the metabolism of PAO (like the specific anaerobic P-release and anaerobic carbon-uptake rates) are not reported (Panswad and Anan, 1999; Intrasugkha *et al.*, 1999; Uygur and Kargi, 2004; Kargi and Uygur, 2005; Uygur, 2006; Hong *et al.*, 2007; Cui et al., 2009). In most cases, only the phosphorus removal efficiency was measured and just a few studies aimed to determine the impact of salinity on the process kinetic rates (Uygur, 2006; Wu *et al.*, 2008). Furthermore, no studies have been undertaken yet on enriched cultures of PAO and GAO (the competing microorganisms in an EBPR system). Overall, it still remains unclear how salinity affects the EBPR process and the microbial populations involved in or related to the process (like PAO and GAO, respectively).

To get a better understanding of the impact of salinity on EBPR, we investigated the effects of salinity on the anaerobic metabolism (stoichiometry and kinetics) of an enriched PAO II and GAO culture, as the competition between PAO and GAO occurs in the anaerobic stage of the process. Enriched PAO and GAO cultures were cultivated in sequencing batch reactors (SBR). Through the execution of short-term (hours) anaerobic batch tests, the effects of salinity on the anaerobic physiology of PAO and GAO are studied with the aim to assess the competition between these two groups of microorganisms under saline conditions and consequently assess the feasibility to apply the EBPR process for the treatment of saline wastewaters.

6.2 Material and methods

6.2.1 Enrichment of PAO and GAO cultures

6.2.1.1 Operation of SBR

PAO and GAO cultures were enriched in two similar double-jacketed laboratory sequencing batch reactors (SBR). The reactors were operated and controlled automatically in a sequential mode (SBR) by an Applikon ADI controller. Online operating data (e.g. pH and O_2) was stored using the BioXpert software (Applikon, The Netherlands, Schiedam). Each SBR had a working volume of 2.5 L. Activated sludge from a municipal wastewater treatment plant with a 5-stage Bardenpho configuration (Hoek van Holland, The Netherlands) was used as inoculum for the enrichment of the PAO culture, while enriched sludge from the PAO SBR (PAO-SBR) was used as inoculum for the GAO SBR (GAO-SBR).

The SBRs were operated in cycles of 6 hours (2.25 h anaerobic, 2.25 aerobic and 1.5 settling phase) following similar operating conditions used in previous studies (Smolders *et al.*, 1994a; Brdjanovic *et al.*, 1997). pH was maintained at 7.2 ± 0.05 (PAO-SBR) and 7.0 ± 0.05 (GAO-SBR) by dosing 0.4 M HCl and 0.4 M NaOH. Temperature was controlled in both reactors at 20 ± 1 °C.

The cycle started with nitrogen gas sparging for 5 minutes at a flow rate of 30 L/h to remove any oxygen remaining from the previous cycle and feed the substrate to the SBR under truly anaerobic conditions. After the first 5 minutes, 1.25 L of synthetic substrate was fed to the SBR over a period of 5 minutes. In the aerobic phase, compressed air was introduced to the SBR at a flow rate of 60 L/h. The Resulting DO concentrations in the PAO-SBR were ranging between 6.4 and 8.2 mg O_2/L and in the GAO-SBR between 2.7 and 8.2 mg O_2/L.

The PAO-SBR and GAO-SBR were controlled at a nominal sludge retention time (SRT) of 8 and 10 days, respectively. For the estimation of the applied SRT, any potential loss of biomass, through the effluent or from the removal of the biofilms stuck to the walls of the reactors during the maintenance of the systems, was not considered. At the end of the settling period, supernatant was pumped out from the reactors, leaving 1.25 L of mixed liquor in the reactor resulting in a total hydraulic retention time (HRT) of 12 h. The mixed liquor was mixed by impellers operated at 500 rpm and 300 rpm for the PAO-SBR and GAO-SBR, respectively, except during settling and decant phases when mixing was switched off.

6.2.1.2.Synthetic substrate

The main difference between the synthetic substrate supplied to the PAO-SBR and GAO-SBR was the phosphorus content. Namely, the influent of PAO-SBR contained 20 mg PO_4^{3-}-P/L (0.65 P-mmol/L), leading to an influent P/C ratio of 0.05 (P-mol/C-mol), while the P content in GAO-SBR influent was limited to 2.2 mg PO_4^{3-}-P/L (0.07 P-mmol/L) (Liu *et al.*, 1997), resulting in an influent P/C ratio of 0.005 (P-mol/C-mol).

The concentrated medium was prepared in two separate solutions with demineralised water. The first solution contained only acetate (HAc) as carbon source, while the other contained all other nutrients, minerals and trace elements required for biomass cultivation. After mixing the two solutions (prior to addition to the respective SBR), the influent of the reactors contained per litre: 860 mg NaAc·3H$_2$O (12.6 C-mmol/L, 405 mg COD/L), 107 mg NH$_4$Cl (2 N-mmol/L), 140 mg CaCl$_2$.2H$_2$O, 2 mg/L of N-Allylthiourea (ATU) to inhibit nitrification, 0.3 mL/L trace elements solution, 89 mg NaH2PO4.H2O

(0.65 P-mmol/L, 20 mg PO_4^{3-}-P/L) for the PAO-SBR, whereas the GAO-SBR contained 9.8 mg $NaH_2PO_4.H_2O$, (2.2 mg PO_4^{3-}-P/L, 0.07 P-mmol/L), 120 mg $MgSO_4.7H_2O$ for PAO-SBR and 90 mg $MgSO_4.7H_2O$ for GAO-SBR, 480 mg KCl for PAO-SBR and 360 mg KCl for GAO-SBR. The trace element solution was prepared as described by Smolders *et al.*, (1994a). Prior to use, both concentrated solutions were autoclaved at 110 $^{\circ}$C for 1 h.

6.2.1.3 Monitoring of the sludge performance

The performance of PAO-SBR and GAO-SBR was regularly monitored by measuring orthophosphate (PO_4^{3-}-P), acetate (Ac-C), mixed liquor suspended solids (MLSS) and mixed liquor volatile suspended solids (MLVSS). The pseudo steady-state conditions of the reactors were confirmed based on daily observations of the aforementioned parameters as well as online pH and DO profiles recorded with the BioXpert software. When no significant changes of these parameters were observed on the online profiles for a time interval of at least 3 SRT, the condition was considered to be in pseudo steady-state.

Cycle measurements were carried out to determine the biomass activity when both SBR reached steady-state conditions. In the cycle measurements, polyhydroxyalkanoate (PHA) and glycogen were also measured in addition to (PO_4^{3-}-P), (Ac-C), (MLSS) and (MLVSS). Estimation of the degree of enrichment of the bacterial populations of interest (PAO and GAO) was undertaken via FISH analysis.

6.2.2 Anaerobic batch tests

After the biomass activity reached pseudo steady-state conditions in the SBR, anaerobic batch experiments at different salt concentrations were performed in two double-jacketed laboratory reactors with a maximal operating volume of 0.5 L. In order to conduct anaerobic short-term salinity tests, a defined volume of enriched PAO and GAO sludge was withdrawn at the end of the aerobic phase from the PAO-SBR and GAO-SBR, respectively, and transferred to the 0.5 L batch reactor. After each sludge transfer, the wastage of sludge was adjusted to compensate for the sludge withdrawal due to batch tests and keep a stable SRT in the parent SBR. Batch tests were performed at controlled temperature and pH (20 ± 0.5 $^{\circ}$C and 7.0 ± 0.05, respectively). pH was automatically maintained by dosing of 0.1 M HCl and 0.1 M NaOH. During the anaerobic batch experiments, the sludge was constantly stirred at 300 rpm.

The batch tests carried out for the determination of the anaerobic kinetic rates of PAO and GAO were executed after the addition of the same synthetic media used for their cultivation but, in addition, NaCl was added at certain defined concentrations. Prior to addition of the synthetic medium in the batch reactor, there was no settling and effluent removal phase, which normally took place in the SBR before adding the influent in the beginning of the next cycle. To execute the anaerobic batch tests with the same acetate-to-biomass ratio as in the parent PAO-SBR and GAO-SBR, the concentration of HAc was reduced by 50%. To suppress any potential foam formation, a drop of silicon antifoam (1% solution) was also added. All batch tests were carried out for 3h. During this period, samples for the determination of MLVSS, MLSS, glycogen, PO_4 and acetate samples were taken. The salinity concentrations studied were chosen in a step-wise manner from 0 to 4% salinity (w/v) to cover the whole inhibition range of the microorganisms (from 0 to 100% relative inhibition). The salinity concentrations assessed for PAO were 0.02, 0.27, 0.52, 0.60, 0.81, 1.02, 2.02, 4.02 % (W/V) NaCl and for GAO they were 0.02, 0.14, 0.27, 0.39, 0.52, 0.77, 1.02, 1.52, 2.02, 3.02, 4.02 % (W/V) NaCl.

6.2.3 Anaerobic maintenance tests

Anaerobic batch tests, similar to those described previously, but without acetate addition, were conducted for the determination of the anaerobic maintenance requirements of PAO and GAO following a procedure described elsewhere (Brdjanovic et al., 1997; Lopez-Vazquez *et al.*, 2007).

N_2 gas was sparged at a flow rate of 6 L/h into the mixed liquor for 5 minutes before the beginning and during the first 15 min of the batch tests to create anaerobic conditions. Thereafter, N_2 was flushed through the headspace of the reactor to prevent oxygen intrusion. Influent was also sparged with N_2 for 15 minutes to remove any potential dissolved oxygen present. After influent addition, the activity of PAO was followed for 3h, and that of GAO for 8h. During this period, samples for the determination of the orthophosphate concentrations were collected for PAO, while in the case of GAO, samples for glycogen analysis were collected. For PAO, two additional anaerobic maintenance tests were conducted to check if a potential shift from poly-P to glycogen consumption occurred. The tests were similar to those described for GAO. The salinity concentrations studied in the PAO and GAO anaerobic maintenance tests were in the range of 0 to 6% NaCl (w/v) for both organisms. Similar to the HAc uptake tests, these salinity concentrations were selected to cover the whole inhibition range, which was higher in the maintenance tests (up to 6% NaCl) than in the HAc uptake tests (up to 4% NaCl). The salinity concentrations assessed for PAO were 0.02, 0.27, 0.52, 1.02, 1.52, 2.02, 2.52, 3.02, 3.52, 4.02, 6.02 % (W/V) NaCl and for GAO they were 0.02, 1.02, 2.02, 3.02, 4.02, 5.02, 6.02 % (W/V) NaCl.

6.2.4 Anaerobic kinetic and stoichiometric parameters

6.2.4.1 PAO and GAO kinetics

The active biomass concentratoins were determined, following the procedures described in section 2.2.4 in chapter 2. All rates were expressed as maximum active biomass specific rates based on the PO_4 and HAc profiles observed in the tests as described by Smolders *et al.* (1994b), Zeng *et al.* (2003), Brdjanovic et al., 1997) and Lopez-Vazquez *et al.*, 2007. For PAO, the HAc uptake rates were corrected for the presence of GAO by subtracting from the total observed HAc uptake rates the HAc uptake rate of GAO and dividing this value by the PAO fraction in the sludge from the PAO-SBR. The anaerobic P-release rates of PAO were also corrected for the presence of GAO by dividing the total P-release rates by the PAO fraction, determined by quantitative FISH analysis, and for the anaerobic maintenance activity by subtracting the maintenance P-release from the total anaerobic P-release.

In anaerobic tests executed without HAc, the effect of salinity on the maintenance activity was assessed by measuring the P-release and glycogen consumption of PAO and GAO. For PAO the maintenance activity was corrected for the presence of GAO in the PAO-SBR sludge. The PAO maintenance activity was determined by dividing the observed P-release rates in the batch tests by the fraction of PAO in the sludge. The anaerobic specific ATP maintenance coefficients of PAO and GAO were determined by linear regression of the glycogen consumption and P-release profiles as described by Smolders *et al.* (1994a) and Zeng *et al.* (2003), respectively. However, when the PAO and GAO glycogen consumption profiles were leveling off over time due to a progressive inhibition at high salinity concentrations, only the initial glycogen consumption observed during the first 2 hours of the tests were considered for the determination of the anaerobic maintenance coefficient.

6.2.4.2 PAO and GAO Stoichiometry

Total P-release rate per HAc uptake rate and corrected P-release rate per HAc uptake rate and total glycogen consumption per HAc uptake were the stoichiometric parameters assessed for PAO. For GAO, the total glycogen consumption per HAc uptake, and corrected glycogen consumption per HAc uptake were the stoichiometric parameters evaluated. The corrected glycogen consumption of PAO and GAO was determined by subtracting the estimated glycogen consumption due to maintenance from the total glycogen consumption observed in the batch tests with HAc.

6.2.5 Analyses

PO_4^{3-}-P (determined by the ascorbic acid method), MLSS and MLVSS determination were performed in accordance with Standard Methods (A.P.H.A., 1995). HAc was determined using a Varian 430-GC Gas Chromatograph (GC), equipped with a split injector (split ratio 1:10), a WCOT Fused Silica column with a FFAP-CB coating (25 m x 0.53mm x 1μm), and coupled to a FID detector. Helium gas was used as carrier gas. Temperature of the injector, column and detector were 200°C, 105°C and 300°C, respectively. PHA content (as PHB and PHV) of freeze dried biomass was determined according to the method described by Smolders *et al.* (1994a). Glycogen determination was also executed according to the method described by Smolders *et al.* (1994b) but with a digestion phase extended to 5 h.

To determine the microbial population distribution of the enriched PAO and GAO culture, Fluorescence *in situ* Hybridization (FISH) was performed according to the procedures described in Amman (1995). In order to target the entire bacterial population, the EUBMIX probe (mixture of probes EUB 338, EUB338-II and EUB338-III) was used. *Accumulibacter* was targeted by a PAOMIX probe (mixture of probes PAO462, PAO651 and PAO846) (Crocetti *et al.*, 2000) whereas a GAOMIX probe (mixture of probes GAOQ431 and GAOQ989) (Crocetti *et al.*, 2002) was used to target *Competibacter*. In order to distinguish the different PAO clades, PAO I (clade 1A and other type 1 clades) and PAO II (clade 2A, 2C and 2D) were targeted by the probes Acc-1-444 and Acc-2-444 (Flowers et al., 2009), respectively. The FISH samples were hybridized under the conditions described by Crocetti *et al.* (2000, 2002) and Flowers *et al.* (2009).

The quantification of the PAO and GAO fractions in the biomass from PAO-SBR was carried out using the free ImageJ software package (version1.47b, Wayne Rasband, National Institute of Health, USA). 8-bit images for each of the color channels (red for PAO, Cy3; green for GAO, Fluos; blue for EUB, Cy5) were converted into binary format. Image coverage was computed by dividing the number of pixels corresponding to the object with the total number of pixels of the image. Fractions of PAO and GAO were calculated as the ratio between their image coverage and that of the entire bacterial population. Around twenty separate images were evaluated.

6.2.6 Parameter fitting and evaluation of models

A structured model was developed (Appendix 6A-C), describing the salinity effects on the different metabolic processes (Equation 1 - 9), to obtain better insight in the salinity effects on the kinetic rates and stoichiometry and to propose a model for practical applications. A detailed explanation of the equations is given in the Appendix 6.A-C.

PAO and GAO Anaerobic Maintenance Coefficient

$$m_{ATP,PAO_{poly}-P}^{an}(S) = (m_{ATP}^0 + a.S).\frac{1}{1+e^{(bi_1.(S-Si_1))}} \qquad \text{(Equation 1)}$$

$$m_{ATP,GAO}^{an}(S) = (m_{ATP}^0 + a.S).\frac{1}{1+e^{(bi_1.(S-Si_1))}} \qquad \text{(Equation 2)}$$

Being,

$m_{ATP,PAO_{poly}-P}^{an}(S)$: PAO poly-P maintenance coefficient at different salinity concentrations

$m_{ATP,GAO}^{an}(S)$: GAO maintenance coefficient at different salinity concentrations

S: Salinity concentration

And fitted parameters,

m_{ATP}^0 : Maintenance coefficient at 0% salinity concentration

a: Linear proportional increase in maintenance requirements per increase in salinity

bi_1: Impact factor, describing the magnitude of the inhibition effect

Si_1: Salinity concentration at which 50% inhibition occurs

PAO and GAO Anaerobic kinetic rates

$$q_{SA,PAO\_total}^{MAX}(S) = q_{SA,PAO\_PAM}^{MAX,0} * \frac{1}{1+e^{(bi_2.(S-Si_2))}} + q_{SA,PAO\_GAM}^{MAX,0} * \frac{1}{1+e^{(ba_1.(Sa_1-S))}} * \frac{1}{1+e^{(bi_1.(S-Si_1))}}$$

(Equation 3)

$$q_{SA,GAO}^{MAX}(S) = q_{SA,GAO}^{MAX,0} * \frac{1}{1+e^{(bi_2.(S-Si_2))}}$$

(Equation 4)

$$q_{P,PAO\_HAc}^{MAX}(S) = q_{P,PAO\_HAc}^{MAX,0} * \frac{1}{1+e^{(bi_2.(S-Si_2))}}$$

(Equation 5)

Being,

$q_{SA,PAO\_total}^{MAX}(S)$: Total maximum PAO acetate uptake rate at different salinity

$q_{SA,GAO}^{MAX}(S)$: Maximum GAO acetate uptake rate at different salinity

$q_{P,PAO\_HAc}^{MAX}(S)$: Maximum PAO PO$_4$ release rate at different salinity

S: Salinity concentration

And fitted parameters,

$q_{SA,PAO\_PAM}^{MAX,0}$: Maximum PAO acetate uptake rate facilitated by a PAO metabolism (PAM) at 0% salinity

$q_{SA,PAO\_GAM}^{MAX,0}$: Maximum PAO acetate uptake rate facilitated by a GAO metabolism (GAM) at 0% salinity

$q_{SA,GAO}^{MAX,0}$: Maximum GAO acetate uptake rate at 0% salinity

$q_{P,PAO\_HAc}^{MAX,0}$: Maximum PAO PO_4 release rate at 0% salinity

bi_2: Impact factor, describing the magnitude of the inhibition effect on the acetate uptake and P-release effect

Si_2: Salinity concentration at which 50% inhibition of the acetate uptake occurs

ba_1: Impact factor (equal to bi_2), describing the magnitude of the activation effect on the acetate uptake

Sa_1: Salinity concentration (equal to Si_2) at which 50% activation of the PAO GAM acetate uptake occurs.

PAO and GAO Anaerobic Stoichiometry

$$f_{P/HAc}^{total}(S) = \frac{q_{P,HAc}^{MAX}(S) + m_{ATP}^{an,re}(S)}{q_{SA,PAO\_total}^{MAX}(S)} \qquad \text{(Equation 6)}$$

$$f_{P/HAc}^{HAc}(S) = \frac{q_{P,HAc}^{MAX}(S)}{q_{SA,PAO\_total}^{MAX}(S)} \qquad \text{(Equation 7)}$$

$$f_{\Delta gly/\Delta HAc}^{total} = \frac{\Delta HAc * f_{\Delta gly/\Delta HAc}^{HAc} + \Delta t * m_{ATP}^{an}(S) * X_{GAO} * 2}{\Delta HAc} \qquad \text{(Equation 8)}$$

$$f_{\Delta gly/\Delta HAc}^{HAc} = fixed\ stoichiometric\ value \qquad \text{(Equation 9)}$$

being,

$f_{P/HAc}^{total}$: total P-release rate /HAc uptake rate at different salinity

$f_{P/HAc}^{HAc}$: P-release rate corrected for maintenance acitvity/HAc uptake rate at different salinity

$f_{\Delta gly/\Delta HAc}^{total}$: net glycogen consumption / net HAc uptake

$f_{\Delta gly/\Delta HAc}^{HAc}$: net glycogen consumption corrected for maintenance acitvity / net HAc uptake

ΔHAc: net acetate consumption during the tests

Δt: time interval of batch test

X_{GAO}: active biomass concentration of GAO

S: Salinity concentration

Model parameters were fitted with the experimental values of the kinetic rates and stoichiometry at different salinity concentrations using the method of least squares and the model simulation was evaluated using the ordinary least squares regression model (Mesple et al., 1996).

6.3 Results

6.3.1 Enrichment of PAO

The PAO-SBR was continuously operated for over 400 days. When it reached pseudo steady-state conditions, the biomass exhibited complete acetate uptake in the anaerobic stage and complete P-removal in the aerobic phase. The MLSS and MLVSS concentration in the reactor was 2913 and 1990 mg/L, respectively, leading to a MLVSS/MLSS ratio of 0.69. This indicates a high ash content in the cell, as usually observed in enriched PAO cultures as a consequence of the high amount of the stored poly-phosphate. During the steady-state period, the average observed P-release/HAc-uptake ratio was 0.33 (s.d. = 0.02, n=18)P-mol/C-mol.

The change in carbon storage compounds and soluble phosphate in a steady state SBR cycle is shown in Figure 6.1a and the composition of the microbial community by FISH is shown in Figure 6,1b. The observed specific acetate uptake rate was 0.25 C-mol/(C-mol.h), the specific anaerobic phosphate release rate was 0.085 P-mol/(C-mol.h) and the specific aerobic phosphate uptake rate was 0.081 P-mol/(C-mol.h). FISH images show that the PAO-SBR sludge was indeed dominated by PAO, although a small fraction of GAO was present. *Accumulibacter* Type II was the dominant microorganism whereas *Accumulibacter* Type I was not observed. Quantitative FISH analysis indicated that PAO and GAO together comprised 98% of the microbial community. The fractions of PAO and GAO were 94% (s.d. = 4%, n=20) and 5% (s.d. = 4%, n=20), respectively.

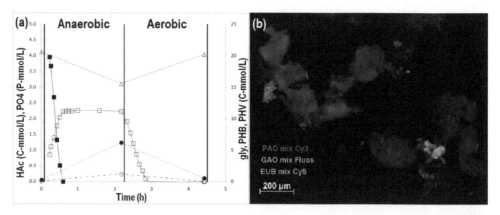

Figure 6.1 Enrichment of the PAO culture: (a) Profile observed during a cycle under pseudo steady-state conditions in the PAO-SBR: acetate (■), PO₄ (□), PHV (○), PHB (•), and glycogen (△) concentrations. (b) FISH image obtained from PAO-SBR: PAO mix (purple), GAO mix (cyan green), EUB (blue)

6.3.2 Enrichment of GAO

The GAO-SBR was continuously operated for more than 350 days. Once the biomass activity in the reactor was stable and reached (pseudo-) steady-state conditions, the MLSS and MLVSS concentrations in the GAO-SBR were 2193 and 2067 mg/L, respectively. The calculated MLVSS/MLSS ratio was 0.94, indicating that no significant polyphosphate accumulation occurred. Figure 6.2 displays a cycle illustrating the biomass activity observed in the GAO-SBR (Figure 6.2a). The FISH analyses (Figure 6.2b) confirmed that the sludge was dominated by GAO (*Competibacter*) and that only minor traces of PAO were present in the sludge. Since FISH quantification would not add significant additional information about the enrichment of the GAO sludge, it was not conducted. The specific acetate uptake rate was 0.18 C-mol/(C-mol.h) and the P-release/HAc-uptake ratio was 0.012 P-mol/C-mol.

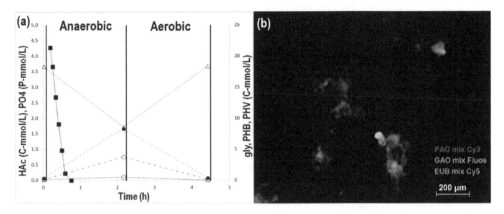

Figure 6.2 Enrichment of the GAO culture: a) Profile observed during a cycle under pseudo steady-state conditions in the GAO-SBR: acetate (■), PO₄ (□), PHV (○), PHB (●), Glycogen (Δ). b) Fish image obtained from GAO sludge: PAO mix (purple), GAO mix (cyan green), EUB (blue)

6.3.3 Impact of salinity on the anaerobic kinetics of PAO and GAO

6.3.3.1 *PAO anaerobic maintenance requirements*
At the different salinity concentrations, all anaerobic maintenance P-release profiles were linear, indicating that no progressive inhibition took place over time during the execution of each batch activity test (data not shown). On the basis of the P-release profiles, from 0 to 1.5% salinity, a continuous increase in the anaerobic maintenance ATP coefficient of PAO was observed, whereas above 2% salinity it decreased (Figure 6.3a). To check if the observed decrease in P-release was compensated by glycogen consumption, two additional tests were carried out at 1 and 4% salinity. In these tests, glycogen consumption was also observed. At 1% salinity, the initial rate (determined during the first 2 h of the test) of maintenance energy production by glycogen conversion (m^{an}_{ATP,PAO_gly}) was 1.6 times higher than the maintenance energy production rate by P-release (m^{an}_{ATP,PAO_poly-P}), while at 4% the m^{an}_{ATP,PAO_gly} was 5.0 times higher than the m^{an}_{ATP,PAO_poly-P}. This indicates that a shift in the energy generating mechanism for maintenance took place, from a poly-P and glycogen dependent mechanism to a mechanism mainly driven by glycogen. The total maintenance energy, produced by glycogen conversion and P-release (m^{an}_{ATP,PAO_total}), at 1% and 4% salinity were 0.0065 and 0.015 ATP-mol/(C-mol.h), respectively. Concerning the glycogen consumption profiles observed at 1 and 4% salinity, it was linear at 1%; however, at 4% salinity it slowed down and leveled off during the test. This suggests that the anaerobic maintenance mechanism of PAO dependent on glycogen was progressively inhibited at 4% salinity.

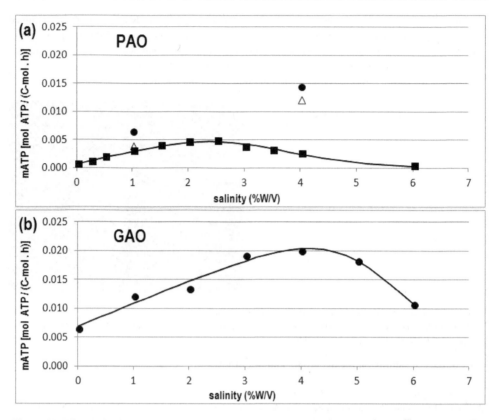

Figure 6.3 Effects of salinity on the anaerobic maintenance coefficients of (a) PAO: m^{an}_{ATP,PAO_poly-P} (■), m^{an}_{ATP,PAO_gly} (△), m^{an}_{ATP,PAO_total} (●), simulation of m^{an}_{ATP,PAO_poly-P} (solid line), and b) GAO: m^{an}_{ATP} (●),simulation of m^{an}_{ATP} (solid line)

6.3.3.2 GAO anaerobic maintenance requirements

In all anaerobic maintenance tests, at increased salinity concentrations the anaerobic glycogen consumption increased, but full glycogen depletion was not observed within the test duration of 8 hours (data not shown). Similar to PAO but at salinity concentrations above 2%, the glycogen consumption profiles leveled off during the tests, implying that GAO were increasingly inhibited over time. In spite of this effect, the initial maintenance energy production rate by glycogen conversion (determined during the first 2 h of the test) of GAO ($m^{an}_{ATP,GAO}$) increased up to 4% salinity (Figure 6.3b), but above 4% it decreased.

6.3.3.3 Modeling the anaerobic maintenance of PAO and GAO

The PAO m^{an}_{ATP,PAO_poly-P} and GAO $m^{an}_{ATP,GAO}$ anaerobic maintenance activities at different salinity levels were successfully described by the empirical Equation 1 and 2. For PAO, both glycogen consumption and P-release should be considered for a complete description of the anaerobic maintenance coefficient m^{an}_{ATP,PAO_total} at different salinities, but the glycogen data at different salinities were insufficient.

Based on the development of Equation 1 and 2 (Appendix 6A), the parameter values of these equations are shown in Table 6.1.

Table 6.1 Values of the coefficients and parameters used for the description of PAO and GAO maintenance activities at different salinity concentrations for Equation 1 and 2.

Kinetics	Parameter	Unit	$m^{an}_{ATP\_poly-P}$	$m^{an}_{ATP,GAO}$
Maintenance rate	$m^0_{ATP}{}^a$	[ATP-mol/(C-mol.h)]	0.0007	0.0069
	a^a	[ATP-mol/(C-mol.h.(%W/V))]	0.0025	0.0039
	bi_1	[1/%(W/V)]	1.32	1.39
	Si_1	(% W/V)	3.11	5.56
Evaluation of maintenance rates simulation	RSQ		0.981	0.970
	Slope	N.A.	0.988	1.001
	intercept		0.000	0.000

^{a}: Parameters obtained by linear regression analysis of the first 4 data points from the anaerobic maintenance tests.

At 0% salinity, the m^0_{ATP} for GAO was 10-fold higher than that of PAO (m^0_{ATP}). Also, the increase in energy maintenance requirements per increase in salinity (a) was higher for GAO. 50% inhibition of the m^{an}_{ATP,PAO_poly-P} and $m^{an}_{ATP,GAO}$ of GAO occurred at 3.1 and 5.6 % salinity for PAO and GAO, respectively. However, the m^{an}_{ATP,PAO_gly} at 4% salinity shows that the glycogen consumption is less inhibited and still occurs at higher salinity concentrations (Figure 6.3a).

6.3.3.4 Maximum specific kinetic rates of PAO and GAO

The impact of salinity on the anaerobic acetate uptake of PAO was pronounced and above 2% salinity incomplete acetate uptake was observed within the 3 h duration of the tests (data not shown). Similarly, the anaerobic P-release decreased with increasing salinity (data not shown). The maximum specific acetate uptake and maximum specific effective P-release rates exhibited a drastic decrease (of about 71% and 81%, respectively) when the salinity increased from 0 to 1% (Figure 6.4a). Above 2% salinity, the concentrations of acetate taken up and P released were merely marginal.

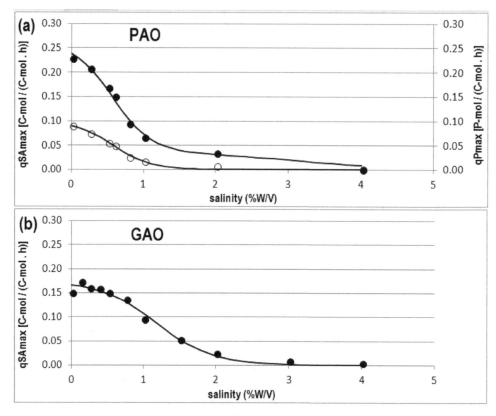

Figure 6.4 Maximum specific kinetic rates of PAO and GAO: (a) PAO HAc uptake rate (●), and PAO PO₄ release rate corrected for maintenance activity (○), during short term tests. b) GAO HAc uptake rate at different salinity concentrations (●). Continuous bold lines show the results of model simulations

For GAO, a gradual decrease in the acetate uptake rate occurred between 0 and 1.5% salinity. At 1% salinity, the maximum specific acetate uptake rate decreased by 41%. Above 1.5% salinity, complete acetate uptake was no longer observed within 3 h (data not shown). 50% inhibition of the maximum specific acetate uptake rates occurred between 1 to 1.5% salinity, further dropping by 85% at 2% salinity (Figure 6.4b).

6.3.3.5 Modeling PAO and GAO kinetics

The impact of salinity on the maximum specific anaerobic HAc uptake and PO₄ release rates of PAO were successfully described by Equations 2 and 4, respectively. Meanwhile, the impact of salinity on the maximum specific anaerobic HAc uptake rate of GAO was successfully described by Equation 3. Based on the development of Equations 2, 3 and 4 (Appendix 6B), the values of their different parameters are shown in Table 6.2.

Table 6.2 Fitted parameters of the equations describing maximum specific HAc uptake rates of PAO and GAO and the maximum specific PO₄ release rate of PAO at different salinities.

Kinetic	Parameter	Unit	PAO	GAO
HAc uptake rates	$q_{SA,PAO\_PAM}^{MAX,0}$	[C-mol/(C-mol.h)]	0.26	N.A.
	$q_{SA,PAO\_GAM}^{MAX,0}$	[C-mol/(C-mol.h)]	0.04	N.A.
	$q_{SA,GAO}^{MAX,0}$	[C-mol/(C-mol.h)]	N.A.	0.18
	$bi_2, ba_1{}^a$	[1/%(W/V)]	3.70	2.55
	$Si_2, Sa_1{}^a$	(%W/V)	0.56	1.18
	$bi_1{}^b$	[1/%(W/V)]	1.32	N.A.
	$Si_1{}^b$	(%W/V)	3.11	
Evaluation of HAc uptake rates simulation	RSQ	N.A.	0.993	0.988
	Slope		0.977	1.017
	Intercept		0.004	0.004
PO₄ release rate^a	$q_{P,HAc}^{MAX,0}$	[P-mol/(C-mol.h)]	0.10	N.A.
	bi_2	[1/%(W/V)]	3.70	
	Si_2	(%W/V)	0.56	
Evaluation of PO₄ release rates simulation	RSQ	N.A.	0.993	
	Slope		1.041	
	Intercept		-0.003	

^{a}: parameters fitted using the PO₄ data
^{b}: parameters are the same as the parameters described in Table 6.1
N.A.: Not Applicable

According to the model parameters, the HAc uptake rate of PAO was 50% inhibited at a salinity concentration of 0.6%, while the HAc uptake rate of GAO at a concentration of 1.2%. Furthermore, the impact factor "bi_2", describing the magnitude of the inhibition effect of salinity on the HAc uptake rate, was higher for PAO than for GAO (3.7 versus 2.6, respectively) reflecting the higher sensitivity of PAO.

6.3.4 Impact of salinity on the anaerobic stoichiometry of PAO and GAO

6.3.4.1 PAO stoichiometry
When the P-release was corrected for the maintenance activity (to display the P-release/HAc uptake ratio caused by HAc uptake only), a significant decrease (of 45%) in the P-release/HAc uptake ratio was observed when the salinity increased up to 2% (Figure 6.5a). If no correction is made, the total P-release/HAc uptake ratio decreased up to 25% at 1% salinity, but at 2% the total P-release/HAc uptake ratio increased due to a higher contribution of P-release for maintenance requirements. Due to limited data collected from the glycogen consumption profiles from the anaerobic maintenance tests, the corrected glycogen/HAc ratio could only be determined at 1% salinity, which was 1.46 C-mol/C-mol. Nevertheless, the total glycogen/HAc ratio (Figure 6.5a) increased from 1.2 to 2.3 C-mol/C-mol when the salinity concentration increased from 0 to 2%.

Table 6.3 Comparison of PAO kinetics and stoichiometry to other values reported in literature

Salinity (%W/V)	PAO fraction (FISH)	GAO fraction (FISH)	SRT days	HRT hours	T °C	pH	m_{ATP}^{an} [mol ATP/ (C-mol.h)]	$q_{SA,total}^{MAX}$ [C-mol / (C-mol.h)]	Total P /HAc (P-mol / C-mol)	Corrected P /HAc (P-mol / C-mol)	Total gly/HAc (C-mol / C-mol)	Corrected Gly/HAc (C-mol / C-mol)	Reference	Comments
0.017	n.d.	n.d.	8	12	20	7	0.0025	0.43	0.52	n.d.	0.5	n.d.	Smolders et al. (1994a)	
0.017	n.d.	n.d.	8	12	20	7	0.0015	0.18	0.38	n.d.	n.d.	n.d.	Brdjanovic et al. (1997)	
0.017	0.85	0.12	10	12	20	7	n.d.	0.17	0.39	n.d.	n.d.	n.d.	Lopez-Vasquez et al. (2007)	
0.017	0.94	0.06	8	12	20	7	0.00071	0.23	0.40	0.40	1.2 (+/- 0.35)	n.d.		Data corrected for the presence of GAO
0.52	0.94	0.06	8	12	20	7	0.0021	0.17	0.34	0.33	1.1 (+/- 0.21)	n.d.	This study	
1.0	0.94	0.06	8	12	20	7	0.0031	0.067	0.30	0.25	0.9 (+/- 0.08)	n.d.		
1.5	0.94	0.06	8	12	20	7	0.0042	n.d.	n.d	n.d.	1.2 (+/- 0.15)	n.d.		
2.0	0.94	0.06	8	12	20	7	0.0047	0.034	0.35	0.22	1.8 (+/- 0.18)	1.5		
4.0	0.94	0.06	8	12	20	7	0.0027	0	n.d	n.d	2.3 (+/- 0.5)	n.d.		
0.016a	n.d.	n.d.	7	8	20	7	0.0024	0.17	n.d.	n.d.	1.2	n.d.	Zeng et al. (2003)	

n.d.: Not determined

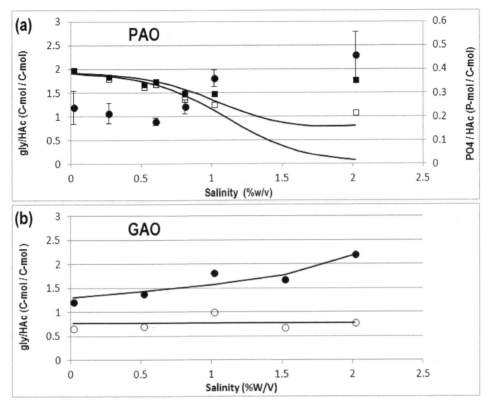

Figure 6.5 Salinity effects on the anaerobic stoichiometry of: (a) PAO: Ratio of total P-release rate/HAc-uptake rate corrected for GAO acetate uptake at different salinities (■), ratio of P-release rate /HAc-uptake rate corrected for maintenance P-release and GAO acetate uptake at different salinities (□), and ratio of total glycogen consumption/HAc-uptake corrected for GAO glycogen consumption at different salinities (●); (b) GAO: ratio of total glycogen consumption /HAc-uptake at different salinity concentrations (●), ratio of glycogen consumption/HAc-uptake corrected for maintenance glycogen consumption (○)

6.3.4.2 GAO stoichiometry

Through the execution of a series of batch tests with an enriched culture of GAO, the effects of salinity on the GAO stoichiometry were assessed. At higher salinity concentrations, the total glycogen consumption/HAc uptake ratio increased significantly (Figure 6.5b). When the stoichiometry was corrected for the maintenance activity (determinedunder the absence of acetate), arguably, the glycogen consumption/HAc-uptake ratio was insensitive to salinity.

6.3.4.3 Modeling PAO and GAO stoichiometry

The experimental values of the anaerobic P-release/HAc-uptake stoichiometry of PAO and gly/HAc-uptake stoichiometry of GAO were predicted (Figure 6.5) with Equation 6 and 7 for PAO and Equation 8 and 9 for GAO developed in Appendix 6C, using the same parameters as described in Table 6.1, Table 6.2 and experimentally determined fixed stoichiometric value of Equation 9 ($f_{\Delta gly/\Delta HAc}^{HAc}$). However, the Equations 6 and 7 were not suitable to describe the PAO stoichiometry above 1% (w/v) salinity. The main reason is that Equation 7 cannot describe the P-release rate accurately above 1% salinity, when P-release is almost fully inhibited. The fixed $f_{\Delta gly/\Delta HAc}^{HAc}$ was determined as the average of the corrected glycogen/HAc-uptake values observed at different salinities with a value of 0.77.

6.4 Discussion

6.4.1 Conversions of PAO and GAO enrichments

The activities of the enriched PAO and GAO cultures at fresh water conditions were compared against those from literature. Concerning the activity of PAO (Table 6.3), the m_{ATP,PAO_poly-P}^{an}, determined from the P-release profiles at 0% salinity are lower than those reported by Smolders *et al.* (1994a) and Brdjanovic *et al.* (1997). In the original experimental design of this study, P-release was the only mechanism considered for energy production during anaerobic starvation. However, in the two additional tests at 1% and 4% salinity, PAO were capable of using glycogen under anaerobic starvation conditions for maintenance purposes, which was also observed in previous studies by Lopez et al., (2006) and Lu et al., (2007). Potentially, in this study at 0% salinity, glycogen was also used as another source of energy during anaerobic starvation. Meanwhile, the acetate uptake rates observed under fresh water conditions (Table 6.3) are in the range of those reported in previous studies (Lopez-Vazquez *et al.*, 2007; Brdjanovic *et al.*, 1997; Smolders *et al.*, 1994a).

In the case of GAO (Table 6.4), the maximum acetate uptake rate of GAO obtained at 0% salinity was close to the rates reported by other authors. However, the anaerobic maintenance requirements at 0% salinity determined in this study were higher than those reported previously (Zeng *et al.*, 2003; Filipe *et al.*, 2001; and Lopez-Vazquez *et al.*, 2007). Considering that the enrichment conditions were similar in all the studies compared in Table 6.4, the reasons for the different maintenance requirements cannot be easily explained. As a consequence of the high anaerobic maintenance activity, the corrected glycogen/HAc stoichiometric value (corrected by excluding the glycogen consumption due to maintenance) observed in this study at fresh water conditions was lower when compared to values from previous studies (Zeng et al., 2003; Lopez-Vazquez et al., 2007; and Liu et al., 1994), whereas it is similar to that observed by Filipe et al. (2001). Assuming a theoretical scenario in which the energy demand for HAc transport is zero, the energy demand for the biochemical conversion of acetate into PHA still requires about 1 C-mol glycogen/C-mol HAc or 0.88 C-mol Gly/C-mol HAc, depending on whether the model developed by Zeng et al. (2002) or Filipe et al. (2001) are used. However, the value observed in this study (0.67 C-mol Gly/C-mol HAc) is slightly below these theoretical values, indicating that the glycogen consumption estimated for maintenance requirements might be different during HAc uptake.

6.4.2 Effects of salinity on the kinetics of PAO and GAO

Both PAO and GAO are sensitive to short-term salinity exposure, PAO being comparatively more sensitive. The m_{ATP,PAO_poly-P}^{an} and $m_{ATP,GAO}^{an}$ maintenance activities increased significantly when the salinity increased up to a threshold salinity concentration of 2.5% and 4% for PAO and GAO, respectively. Above this concentration, the GAO $m_{ATP,GAO}^{an}$ and PAO m_{ATP,PAO_poly-P}^{an} decreased while the m_{ATP,PAO_gly}^{an} for PAO continued to increase above 2.5% salinity. Possibly, with the increasing salinity certain salt ions leak through the cell membrane and the organisms needed additional energy to pump out the ions at the expense of energy (Castle *et al.* 1986), which could explain the increasing maintenance activity at higher salinity. The observed decrease in PAO m_{ATP,PAO_poly-P}^{an} maintenance activity above the threshold concentration (2.5% salinity) could be explained by the possible inhibition of the P-release pathway, forcing a shift in the metabolism from poly-P consumption to glycogen consumption, which can be supported by the additional test performed at 4% salinity (Figure 6.3). The non-linear glycogen consumption profiles at high salinity for both PAO and GAO and the decrease in the $m_{ATP,GAO}^{an}$ (above 4% salinity) indicate that above certain threshold concentration the $m_{ATP,GAO}^{an}$ activity and m_{ATP,PAO_gly}^{an} were progressively inhibited. An explanation for this observation could be that at the threshold concentration, the organisms are not able to pump out the ions at the required rate to keep a zero salinity concentration within the intracellular environment. This phenomenon could be caused by the inhibition of the membrane proteins that are responsible for the generation of a proton motive force to pump out the ions or because the maximum rate to expel the ions reaches a limit. Above this threshold concentration, possibly the ions will start to accumulate in the intracellular environment, affecting all metabolic activities, followed by a decrease in all metabolic rates as a consequence of the high intracellular salt concentration as observed in these tests (Figure 6.3).

Concerning the salinity effects on the maximum acetate uptake of PAO and GAO, the salinity concentrations at which 50% inhibition of the HAc-uptake rates took place were 0.6 and 1.2% salinity for PAO and GAO, respectively.When the salinity increased to 1%, PAO showed a 71% decrease in HAc uptake, whereas GAO showed only a 41% decrease. Likely the reduction in activity below 2 or 3% salinity was mostly caused by the inhibition of the acetate uptake enzymes or, in the case of PAO, also due to the inhibition of the P-release enzymes. The decay or lyses of the cells could be discarded because the inhibition observed in the maintenance profiles occurred at higher concentrations.

The observation that the effective P-release rate is nearly zero at 2% salinity, while HAc uptake continues, suggests that a switch in the metabolism occurs from a poly-P dependent PAM HAc uptake to a GAM HAc uptake process that depends on glycogen as sole energy source. Once the salinity concentration is above the maximum threshold concentration (as reported in this study for the maintenance tests), salts are expected to accumulate in the intracellular environment, inhibiting all the intracellular metabolic activities, including all possible pathways for HAc uptake.

6.4.3 Effect of salinity on the stoichiometry of PAO
Between 0 and 2% salinity, a significant decrease (45%) in the corrected P-release/HAc-uptake ratio was observed, which was accompanied by an increase in the total gly/HAc ratio and, at least at 1% salinity, an increase in the corrected glycogen/HAc ratio. The corrected glycogen/HAc at 1% salinity was 1.46 C-mol/C-mol, which was significantly higher than the uncorrected glycogen/HAc ratios between 0 and 0.5% salinity (Table 6.3). As previously suggested, the increase in the corrected glycogen/HAc ratio at 1% salinity and decrease in the corrected P/HAc ratio suggests that a shift from a PAM to a GAM occurred. Zhou et al. (2008), Acevedo et al. (2012) and Welles et al. (2015) demonstrated that PAO are able to switch from a PAM to a GAM when internal poly-P reserves become limited. Thus, the P-release could be completely substituted by glycogen consumption for HAc uptake.

6.4.4 Effect of salinity on the stoichiometry of GAO
Although the total glycogen/HAc stoichiometry increased at higher salinity, the corrected glycogen/HAc stoichiometry, obtained after subtraction of the maintenance requirements, show that the stoichiometry for HAc uptake was not affected by salinity in case of GAO.

6.4.5 Modelling the kinetics and stoichiometry of PAO and GAO
The anaerobic maintenance energy requirements and kinetics of PAO and GAO were successfully described by the proposed model equations. This supports some of the mechanisms proposed in this study, such as the shift in the PAO metabolism from a PAM to a GAM for the uptake of HAc. Although the model could not describe the PAO stoichiometry above 1% salinity, the GAO stoichiometry model described the GAO stoichiometry up to 2%. The proposed model may be used to predict shock-load effects of salt in real wastewater treatment plants, but for such applications, the model should be validated using activated sludge from full scale wastewater treatment plants.

Table 6.4 Comparison of GAO kinetics and stoichiometry to other values reported in literature.

Salinity	SRT	HRT	T	pH	m_{ATP}^{an}	$q_{SA,GAO}^{MAX,total}$	Total P/HAc	Total GLY/HAc	Corrected GLY/HAc	Reference
(%W/V)	(days)	(hours)	°C		[mol ATP/C-mol.h]	[C-mol/(C-mol.h)]	[P-mol/C-mol]	[C-mol/C-mol]	(C-mol/C-mol)	
0.016	7	8	20	7	0.0024	0.17	n.d.	1.2	n.d.	Zeng et al. (2003)
0.016	10	12	20	7	0.0033	0.20	0.01	1.2 (+/-0.19)	n.d.	Lopez-Vasquez et al. (2007)
0.016	7.5-8.0	6	25	7-8	n.d.	n.d.	n.d.	1.1-1.3	n.d.	Liu et al. (1994)
0.016	7	12	22	7.0	0.0021	0.15	n.d.	0.83	n.d.	Filipe et al. (2001)
0.016	8	12	20	7.0	0.0066	0.15	0.012	1.2	0.67	This study
0.52	8	12	20	7.0	n.d.	0.15	n.d.	1.4	0.72	
1	8	12	20	7.0	0.012	0.095	n.d.	1.8	1.0	
1.5	8	12	20	7.0	n.d.	0.052	n.d.	1.7	0.69	
2	8	12	20	7.0	0.014	0.023	n.d.	2.2	0.78	
4	8	12	20	7.0	0.019	0.0039	n.d.	n.d.	n.d.	
0	8	12	20	7	0.0025	0.43	0.52	0.50	n.d.	Smolders et al. (1994a)

n.d.: Not determined.

[a]: Enriched PAO culture

metabolisms followed a similar trend, indicating that both organisms were affected in a similar way. Considering the effect of salinity on the acetate uptake rates of PAO and GAO, the effects on PAO were much more pronounced. Between 0 and 0.6% salinity, PAO showed higher acetate uptake rates under the same operational conditions (pH=7, T=20 °C), whereas GAO showed higher uptake rates above 0.6 % salinity. In theory, the organism with the fastest acetate uptake rate will have a competitive kinetic advantage in the system. Therefore, these results suggest that during short-term exposure to relatively low salinity levels in the range of 0 to 0.6%, PAO may prevail in the system leading to successful EBPR. On the other hand, if the salinity concentrations are above 0.6 % salinity, GAO will likely prevail, leading to deterioration of the EBPR. These observations support the statement that it is feasible to remove phosphorus by biological means when treating saline wastewater with salinity concentrations in a range between 0 to 0.6% (as NaCl).

In order to better understand the effect of salinity on the PAO and GAO cultures at short-term exposure, aerobic batch tests need to be conducted as well. If the aerobic metabolism suffers from serious deleterious effects, the aerobic phase will also play an important role in the occurrence of PAO and GAO as well as on their competition. Ultimately, the most affected part of the metabolism of these organisms (either the anaerobic or aerobic one) will determine the magnitude of their occurrence in the system and the reliability and success of EBPR when treating saline wastewaters. Finally, the assessment of the long-term (months) effects of saline wastewater (NaCl based) on the microbial community competition and physiology of enriched PAO and GAO cultures is necessary to understand the potential selection or adaptation of salt-tolerant strains and to achieve conclusive observations regarding the feasibility to biologically remove phosphorus from saline wastewaters.

6.5 Conclusions

Both PAO and GAO are sensitive to short-term salinity exposure, PAO being comparatively more sensitive. The salinity concentrations at which 50% inhibition of the HAc-uptake rates took place were 0.6 and 1.2% salinity for PAO and GAO, respectively. When the salinity increased to 1%, PAO showed a 71% decrease in HAc uptake, whereas GAO showed only a 41% decrease. The maintenance requirements of both PAO and GAO increased when the salinity increased up to a certain threshold concentration above which the activity ceased. An increase in salinity seemed to induce a larger contribution of glycogen conversion in the anaerobic ATP formation in PAO. Up to 0.6% salinity, PAO exhibited faster uptake rates than GAO under the same operational conditions (pH=7, T=20 °C). Above 0.6% salinity, GAO exhibited higher HAc uptake rates. These results suggest that GAO's might be favoured at higher salinities in wastewater. This needs to be further investigated in future long-term experiments.

6.6 Acknowledgements

This research study was carried out as part of the SALINE project (http://www.salinesanitation.info) led by UNESCO-IHE Institute for Water Education and consortium partners KWR Watercycle Research Institute, Delft University of Technology, University of Cape Town, The Hong Kong University of Science and Technology, The Higher Polytechnic Institute "José Antonio Echeverría" and Birzeit University. The SALINE project is financed by UNESCO-IHE internal research fund with a special generous contribution from Professor George Ekama from University of Cape Town, to whom the authors would like to gratefully thank. Special thanks to UNESCO-IHE laboratory staff, in particular to Don van Galen, for all their support during the research project.

6.7 References

Amann R.I. (1995) In situ identification of microorganisms by whole cell hybridization with rRNA-targeted nucleic acid probes. In: Akkermans ADL, van Elsas JD, de Bruijn FJ, editors. Molecular microbial ecology manual. London: Kluwer Academic Publisher, London, pp 1-15.

Acevedo B., Oehmen A., Carvalho G., Seco A., Borras L., Barat R., (2012) Metabolic shift of polyphosphate-accumulating organisms with different levels of poly-phosphate storage. Water Res 46: 1889-1900

Brdjanovic D., van Loosdrecht M.C.M., Hooijmans C.M., Alaerts G.J., Heijnen J.J. (1997) Temperature effects on physiology of biological phosphorus removal. J Environ Eng-ASCE 123(2): 144-154

Castle A.M. Macnab R.M. Shulman R.G. (1986) Coupling between the Sodium and Proton Gradients in Respiring *Escherichia coli* Cells Measured by 23Na and 31P Nuclear Magnetic Resonance. J Biol Chem 261(17): 7797-7806

Crocetti G. R., Hugenholtz P., Bond P.L., Schuler A., Keller J., Jenkins D., Blackall L.L., (2000) Identification of polyphosphate-accumulating organisms and design of 16S rRNA-directed probes for their detection and quantitation. Appl Environ Microbiol 66: 1175-1182

Crocetti G.R., Banfield J.F., Keller J., Bond P.L., Blackall L.L. (2002) Glycogen accumulating organisms in laboratory-scale and full-scale wastewater treatment processes. Microbiol 148: 3353-3364.

Cui Y., Peng C., Peng Y., Ye L. (2009) Effects of Salt on Microbial Populations and Treatment Performance in Purifying Saline Sewage Using the MUCT Process. Clean-Soil Air Water 37(8): 649-656

Fahim F.A., Fleita D.H. Ibrahim A.M., El-Dars F.M.S. (2000) Evaluation of some methods for fish canning wastewater treatment. Water Air Soil Poll 127: 205-226

Filipe C.D.M., Daigger G.T., Grady C.P.L. (2001) A metabolic model for acetate uptake under anaerobic conditions by glycogen-accumulating organisms: stoichiometry, kinetics and effect of pH. Biotechnol Bioeng 76(1): 17-31.

Flowers J.J., He S., Yilmaz S., Noguera D.R., McMahon K.D. (2009) Denitrification capabilities of two biological phosphorus removal sludges dominated by different '*Candidatus* Accumulibacter' clades. Environ Microbiol Rep (2009) 1(6): 583-588

Gonzalez J.F., Civit E.M., Lupin H.M. (1983) Composition of fish filleting wastewater. Water SA 9(2): 49-56

Hong C.C., Chan S.K., Shim H. (2007) Effect of chloride on biological nutrient removal from wastewater. J Appl Sci Environ Sanit 2(3): 85-92

Intrasungkha N., Keller J., Blackall L.L. (1999) Biological nutrient removal efficiency in treatment of saline wastewater. Water Sci Technol 39(6): 183-190

Kargi, F., A. Uygur, (2005), Improved Nutrient Removal from Saline Wastewater in an SBR by *Halobacter* Supplemented Activated Sludge. Envir Eng Sci 22 (2): 170-176

Lefebvre O., Moletta R., (2006) Treatment of organic pollution in industrial saline wastewater: A literature review. Water Res40: 3671-3682.

Leung R.W.K., Li D.C.H., Yu W.K., Chui H.K., Lee T.O., Van Loosdrecht M.C.M., Chen G.H. (2012) Integration of seawater and grey water reuse to maximize alternative water resource for coastal areas: The case of the Hong Kong International Airport. Water Sci Technol 65(3): 410-417

Liu W.T., Mino T., Nakamura K., Matsuo T. (1994) Role of glycogen in acetate uptake and polyhydroxyalkanoate synthesis in anaerobic-aerobic activated sludge with a minimized polyphosphate content. J Ferment Bioeng 77(5): 535-540

Liu W.T., Nakamura K., Matsuo T., Mino T. (1997) Internal energy-based competition between poly-phosphate- and glycogen-accumulating bacteria in biological phosphorus removal reactors-effect of P/C feeding ratio. Water Res 31(6): 1430-1438

Lopez-Vazquez C.M., Song Y.I., Hooijmans C.M., Brdjanovic D., Moussa M.S., Gijzen H.J., van Loosdrecht M.C.M. (2007) Short-term temperature effect on the anaerobic metabolism of Glycogen Accumulating Organisms. Biotechnol Bioeng 97(3): 483-495

Lopez C., Pons M.N. Morgenroth E. (2006) Endogenous processes during long-term starvation in activated sludge performing enhanced biological phosphorus removal. Water Res 40(8): 1519-1530

Lu H., Keller J., Yuan Z. (2007) Endogenous metabolism of Candidatus Accumulibacter phosphatis under various starvation conditions. Water Res41(20): 4646-4656

Mesple F., Troussellier M., Casellas C., Legendre P. (1996) Evaluation of simple statistical criteria to qualify a simulation. Ecol Model 88(1-3): 9-18

Orhon D., Tasli R., Sozen S. (1999) Experimental basis of activated sludge treatment for industrial wastewaters - the state of the art. Water Sci Technol 40(1): 1-11

Panswad T., Anan C. (1999) Impact of high chloride wastewater on an anaerobic/anoxic/aerobic process with and without inoculation of chloride acclimated seeds. Water Res33(5): 1165-1172

Smolders G.J.F., Van der Meij J., Van Loosdrecht M.C.M. and Heijnen J.J. (1994a) Model of the anaerobic metabolism of the biological phosphorus removal process: Stoichiometry and pH influence. Biotechnol Bioeng 43: 461-470

Smolders G.J.F., van der Meij J., Van Loosdrecht M.C.M. and Heijnen J.J. (1994b) Stoichiometric model of the aerobic biological phosphorus removal process. Biotechnol Bioeng 44:837-848

Tang S.L., Yue D.P.T., Li X.Z. (2006) Comparison of engineering costs of raw freshwater, reclaimed water and seawater for toilet flushing in Hong Kong. Water Environ J 20(4): 240-247

Tian, W.D., Lopez-Vazquez, Li, W.G., C.M., Brdjanovic, Van Loosdrecht, M.C.M. (2013) Occurence of PAOI in a low temperature EBPR system. Chemosphere 92: 1314-1320

Uygur A., Kargi F. (2004) Salt Inhibition on biological nutrient removal from saline wastewater in a sequencing batch reactor. Enzyme Microb Tech, 34: 313-318

Uygur A. (2006) Specific nutrient removal rates in saline wastewater treatment using sequencing batch reactor. Process Biochem 41(1): 61-66

Welles, L., Tian, W. D., Saad, S., Abbas, B., Lopez-Vazquez, C. M., Hooijmans, C. M., van Loosdrecht, M.C.M., Brdjanovic, D. (2015). Accumulibacter clades Type I and II performing kinetically different glycogen-accumulating organisms metabolisms for anaerobic substrate uptake. Water research, 83, 354-366.

WSD (2009) Annual report 2008/2009. Water supply department, Hong Kong SAR Government.

Wu G., Guan Y., Zhan X. (2008) Effect of salinity on the activity, settling and microbial community of activated sludge in a sequencing batch reactors treating synthetic saline wastewater. Water Sci Technol 58(2): 351-358

Zeng R.J., Yuan Z.G., Van Loosdrecht M.C.M., Keller J. (2002) Proposed modifications to metabolic model for glycogen-accumulating organisms under anaerobic conditions Biotechnol Bioeng80 (3): 277–279.

Zeng R.J., van Loosdrecht M.C.M., Yuan Z., Keller J. (2003) Metabolic model for glycogen-accumulating organisms in anaerobic/aerobic activated sludge systems. Biotechnol Bioeng 81(1): 92-105.

Zhou Y., Pijuan M., Zeng R.J., Lu H., Yuan Z., (2008) Could polyphosphate-accumulating organisms (PAO) be glyccogen-accumulating organisms (GAO)? Water Res 42: 2361-2368

6.8 Appendix 6A

Model describing the PAO and GAO Anaerobic Maintenance Coefficient.

While the GAO maintenance energy was generated by glycogen conversion only, the maintenance energy of PAO was produced by both poly-P cleavage and glycogen conversion (Equation 1a). Due to a lack of sufficient glycogen data it was not possible to describe the PAO maintenance energy production by glycogen conversion. Regarding the PAO maintenance energy production by poly-P cleavage and the GAO maintenance energy production by glycogen conversion, it was assumed that the maintenance activity is a product of the maintenance requirements and an empirical inhibition factor (Equation 1b and 2). Due to the need to detoxify, the required maintenance activity increases with increasing salinity (Equation 3). The empirical inhibition factor describes the inhibition of the maintenance activity by the salinity (Equation 4).

$$m_{ATP,PAO\_total}^{an}(S) = m_{ATP,PAO\_poly-P}^{an}(S) + m_{ATP,PAO\_gly}^{an}(S) \qquad \text{(Equation 1a)}$$

$$m_{ATP,PAO\_poly-P}^{an}(S) = m_{ATP}^{an,re}(S).f_{i,1}(S) \qquad \text{(Equation 1b)}$$

$$m_{ATP,GAO}^{an}(S) = m_{ATP}^{an,re}(S).f_{i,1}(S) \qquad \text{(Equation 2)}$$

$$m_{ATP}^{an,re}(S) = (m_{ATP}^{0} + a.S) \qquad \text{(Equation 3)}$$

$$f_{i,1}(S) = \frac{1}{1+e^{(bi_1.(S-Si_1))}} \qquad \text{(Equation 4)}$$

Where:

$m_{ATP,PAO\_total}^{an}(S)$: PAO Maintenance coefficient at different salinity concentrations

$m_{ATP,PAO\_poly-P}^{an}(S)$: PAO poly-P maintenance coefficient at different salinity concentrations

$m_{ATP,PAO\_gly}^{an}(S)$: PAO glycogen maintenance coefficient at different salinity concentrations

$m_{ATP,GAO}^{an}(S)$: GAO Maintenance coefficient at different salinity concentrations

$m_{ATP}^{an,re}(S)$: Required maintenance coefficient at different salinity concentrations

$f_{i,1}(S)$: Empirical inhibition factor at different salinity concentrations

S: Salinity concentration

And fitted parameters,

m_{ATP}^{0} : Maintenance coefficient at 0% salinity concentration

a: Linear proportional increase in maintenance requirements per increase in salinity

bi_1: Impact factor, describing the magnitude of the inhibition effect

Si_1: Salinity concentration at which 50% inhibition occurs

6.9 Appendix 6B.
Model describing the PAO and GAO Anaerobic kinetic rates.

In order to describe the observed profile of the PAO HAc-uptake rates at different salinity concentrations, it was assumed that the PAO total uptake rate of HAc is the sum of a HAc uptake coupled to P-release (PAM) and HAc uptake that solely relies on glycogen consumption (GAM) (Equation 5a). When PAM gets inhibited with increasing salinity (Equation 5b), GAM becomes active (Equation 5c) until a certain salinity concentration above which it becomes inhibited as well. The activation of GAM (Equation 9) was considered to be directly coupled to the inhibition of PAM (Equation 8) and, therefore, it is described by the same parameters as the inhibition of PAM. If the cell is no longer able to generate the required energy for maintenance it was assumed that the whole metabolism would become inhibited (Equation 4). Therefore inhibition of GAM is described by the same inhibition parameters as those used for the inhibition of the maintenance coefficient. The impact of salinity on the anaerobic PO$_4$- release rate of PAO and the anaerobic HAc uptake rate of GAO are described by Equation 6 and 7, respectively.

$$q_{SA,PAO\_total}^{MAX}(S) = q_{SA,PAO\_PAM}^{MAX}(S) + q_{SA,PAO\_GAM}^{MAX}(S) \qquad \text{(Equation 5a)}$$

$$q_{SA,PAO\_PAM}^{MAX}(S) = q_{SA,PAO\_PAM}^{MAX,0} * f_{i,2}(S) \qquad \text{(Equation 5b)}$$

$$q_{SA,PAO\_GAM}^{MAX}(S) = q_{SA,PAO\_GAM}^{MAX,0} * f_{a,1}(S) * f_{i,1}(S) \qquad \text{(Equation 5c)}$$

$$q_{SA,GAO}^{MAX}(S) = q_{SA,GAO}^{MAX,0} * f_{i,2}(S) \qquad \text{(Equation 6)}$$

$$q_{P,PAO\_HAc}^{MAX}(S) = q_{P,HAc}^{MAX,0} * f_{i,2}(S) \qquad \text{(Equation 7)}$$

$$f_{i,2}(S) = \frac{1}{1+e^{(bi_2 \cdot (S - Si_2))}}$$ (Equation 8)

$$f_{a,1}(S) = \frac{1}{1+e^{(ba_1 \cdot (Sa_1 - S))}}$$ (Equation 9)

where,

$q_{SA,PAO\_total}^{MAX}(S)$: Total maximum PAO acetate uptake rate at different salinity

$q_{SA,PAO\_PAM}^{MAX}(S)$: Maximum PAO acetate uptake rate facilitated by a PAM at different salinity

$q_{SA,PAO\_GAM}^{MAX}(S)$: Maximum PAO acetate uptake rate facilitated by a GAM at different salinity

$q_{SA,GAO}^{MAX}(S)$: Maximum GAO acetate uptake rate at different salinity

$q_{P,PAO\_HAc}^{MAX}(S)$: Maximum PAO PO$_4$ release rate at different salinity

$f_{i,2}(S)$: Empirical inhibition factor, describing the inhibition on the PAO acetate uptake and P-release facilitated by a PAM and the GAO acetate uptake at different salinity

$f_{a,1}(S)$: Empirical activation factor, describing the activation of the PAO acetate uptake facilitated by a GAM at different salinity

S: Salinity concentration

And fitted parameters,

$q_{SA,PAO\_PAM}^{MAX,0}$: Maximum PAO acetate uptake rate facilitated by a PAM at 0% salinity

$q_{SA,PAO\_GAM}^{MAX,0}$: Maximum PAO acetate uptake rate facilitated by a GAM at 0% salinity

$q_{SA,GAO}^{MAX,0}$: Maximum GAO acetate uptake rate at 0% salinity

$q_{P,PAO\_HAc}^{MAX,0}$: Maximum PAO PO$_4$ release rate at 0% salinity

bi_2: Impact factor, describing the magnitude of the inhibition effect on the acetate uptake and P-release effect

Si_2: Salinity concentration at which 50% inhibition of the acetate uptake occurs

ba_1: Impact factor (equal to bi_2), describing the magnitude of the activation effect on the acetate uptake

Sa_1: Salinity concentration (equal to Si_2) at which 50% activation of the PAO GAM acetate uptake occurs.

6.10 Appendix 6C.
Model describing the PAO and GAO Anaerobic Stoichiometry.

For PAO, the total P/HAc stoichiometry and corrected P/HAc stoichiometry were described by Equation 10 and 11, which were derived on the basis of Equation 1b, 5 and 7. The total P/HAc stoichiometry (Equation 10) was described as the ratio of the total P-release rate and the HAc-uptake rate (Equation 5). The total P-release rate was described as the sum of the corrected P-release rate (Equation 7) and the anaerobic maintenance coefficient (Equation 1b). The corrected stoichiometry (Equation 11) was described as the ratio of the corrected P-release rate (Equation 7) and the HAc-uptake rate (Equation 5).

$$f_{P/HAc}^{total}(S) = \frac{q_{P,PAO\_HAc}^{MAX}(S) + m_{ATP,PAO_{poly}-P}^{an}(S)}{q_{SA,PAO\_total}^{MAX}(S)} \qquad \text{(Equation 10)}$$

$$f_{P/HAc}^{HAc}(S) = \frac{q_{P,PAO\_HAc}^{MAX}(S)}{q_{SA,PAO\_total}^{MAX}(S)} \qquad \text{(Equation 11)}$$

Where,

$f_{P/HAc}^{total}$: total P-release rate /HAc uptake rate at different salinity

$f_{P/HAc}^{HAc}$: P-release rate corrected for maintenance acitvity/HAc uptake rate at different salinity

For the total stoichiometry and corrected stoichiometry of glycogen consumption/HAc uptake, Equation 12 and 13 were derived on the basis of Equation 1, 5, and 10. The corrected amount of glycogen consumption per amount of HAc-uptake was described by ratio $f_{gly/HAc}^{HAc}$ (Equation 13) as the sum of a fixed stoichiometric value and a compensation factor that compensates the decrease in P/HAc ratio by an ATP equivalent increase in the gly/HAc ratio, which equals the decrease in P/HAc ratio multiplied by a factor 2, as the generation of 1 mol ATP requires 1 P-mol poly-P or 2 C-mol glycogen.

The total $f_{gly/HAc}^{total}(S)$ was described in Equation 12 as the sum of the glycogen consumption rate for HAc uptake and the glycogen consumption for maintenance divided by the HAc uptake rate. The glycogen consumption rate for HAc uptake was described by the HAc-uptake rate multiplied by $f_{gly/HAc}^{HAc}$, while the PAO glycogen maintenance coefficient multiplied by 2 described the the glycogen consumption for maintenance energy production.

$$f_{gly/HAc}^{total}(S) = \frac{f_{gly/HAc}^{HAc} * q_{SA,PAO\_total}^{MAX}(S) + m_{ATP,PAO\_gly}^{an}(S)*2}{q_{SA,PAO\_total}^{MAX}(S)} \qquad \text{(Equation 12)}$$

$$f_{gly/HAc}^{HAc} = fixed\ stoichiometric\ value + 2 * \left(\frac{f_P^{HAc}(0)}{HAc} - \frac{f_P^{HAc}(S)}{HAc} \right) \qquad \text{(Equation 13)}$$

Unfortunately the glycogen data from the PAO maintenance tests were insufficient to verify these equations.

For the GAO total stoichiometry and corrected stoichiometry of glycogen consumption/HAc uptake, equation 14 and 15 were derived on the basis of equation 2 and 6. It was assumed that the amount of glycogen consumption, corrected for maintenance glycogen consumption, per amount of HAc- uptake was a fixed ratio $f_{gly/HAc}^{HAc}$ (Equation 15). The corrected glycogen consumption rate for HAc uptake was therefore described by the HAc-uptake rate multiplied with $f_{gly/HAc}^{HAc}$. The glycogen consumption due to maintenance was described by the GAO maintenance coefficient multiplied by 2, as the generation of 1 mol ATP requires 2 C-mol glycogen. The total glycogen consumption rate could be described by the sum of the corrected glycogen consumption for HAc-uptake rate and glycogen consumption rate for maintenance. Finally, the total stoichiometry was described as the ratio of the total glycogen consumption rate and HAc-uptake rate (Equation 14).

$$f_{gly/HAc}^{total}(S) = \frac{f_{gly/HAc}^{HAc} * q_{SA,GAO}^{MAX}(S) + m_{ATP,GAO}^{an}(S)*2}{q_{SA,GAO}^{MAX}(S)} \qquad \text{(Equation 14)}$$

$$f_{gly/HAc}^{HAc} = fixed\ stoichiometric\ value \qquad \text{(Equation 15)}$$

During the GAO HAc-uptake tests, glycogen samples were only taken at the beginning and end of the tests. Therefore the experimental data was not suitable to verify Equation14 and 15, which represent the ratio of the initial rates. For practical purposes Equation 16 and 17 were derived from Equation14 and 15. The fixed stoichiometric ratio of Equation 17 was assumed to be rate independent and equal to the ratio described in Equation 15. The total net glycogen consumption per net HAc-uptake over the duration of the test was described by Equation 16. The corrected glycogen consumption was coupled to the HAc-uptake similar as described in Equation 17 and the glycogen consumption due to maintenance was described as the product of the activity test duration, the specific maintenance coefficient, the active biomass concentration and the conversion factor 2 (for conversion from ATP to C-mol glycogen).

$$f_{\Delta gly/\Delta HAc}^{total} = \frac{\Delta HAc * f_{\Delta gly/\Delta HAc}^{HAc} + \Delta t * m_{ATP,GAO}^{an}(S)*X_{GAO} * 2}{\Delta HAc} \qquad \text{(Equation 16)}$$

$$f_{\Delta gly/\Delta HAc}^{HAc} = fixed\ stoichiometric\ value \qquad \text{(Equation 17)}$$

Where,

$f_{\Delta gly/\Delta HAc}^{total}$: net glycogen consumption / net HAc uptake

$f_{\Delta gly/\Delta HAc}^{HAc}$: net glycogen consumption corrected for maintenance glycogen consumption / net HAc uptake

$f_{gly/HAc}^{total}$: glycogen consumption rate /HAc uptake rate

$f_{gly/HAc}^{HAc}$: glycogen consumption rate corrected for the maintenance glycogen consumption rate/HAc uptake rate

ΔHAc: net acetate consumption during the tests

Δt: time interval of batch test

X_{GAO}: active biomass concentration of GAO

Impact of salinity on the aerobic metabolism of polyphosphate-accumulating organisms

Abstract

The use of saline water in urban areas for non potable purposes to cope with fresh water scarcity, intrusion of saline water and disposal of industrial saline wastewater into the sewerage lead to elevated salinity levels in wastewaters. Consequently, saline wastewater is generated, which needs to be treated before its discharge into surface water bodies. The objective of this research was to study the effects of salinity on the aerobic metabolism of polyphosphate-accumulating organisms (PAO) which belong to the microbial populations responsible for EBPR in activated sludge systems. In this study, the short-term impact (hours) of salinity (as NaCl) was assessed on the aerobic metabolism of a PAO culture, enriched in a sequencing batch reactor (SBR). All aerobic PAO II metabolic processes were drastically affected by elevated salinity concentrations. The aerobic maintenance energy requirement increased, when the salinity concentration rose up to a threshold concentration of 2% salinity (on a W/V basis as NaCl) while above this concentration the maintenance energy requirements seemed to decrease. All initial rates were affected by salinity, with the NH_4 and PO_4 uptake rates being the most sensitive. A salinity increase from 0 to 0.18% caused a 25%, 46% and 63% inhibition of the O_2, PO_4 and NH_4 uptake rates. The stoichiometric ratios of the aerobic conversions confirmed that growth was the process with the highest inhibition, followed by poly-P and glycogen formation. The study indicates that shock loads of 0.18% salt, which corresponds to the use or intrusion of about 5% seawater may severely affect the EBPR process in wastewater treatment plants not exposed regularly to high salinity concentrations.

Adapted from:

Welles, L., Lopez-Vazquez, C. M., Hooijmans, C. M., van Loosdrecht, M. C. M., Brdjanovic, D. (2015). Impact of salinity on the aerobic metabolism of phosphate-accumulating organisms. *Applied microbiology and biotechnology,99*(8), 3659-3672.

7.1 Introduction

The direct use of saline water (sea and brackish) for non-potable purposes, such as flushing toilets or cooling, seems to be a promising solution to mitigate fresh water stress in urban areas located in coastal zones and inland areas where brackish ground water is available. In comparison to other water production and treatment applications, like desalination or wastewater reclamation, it is a cost-effective and environmentally-friendly alternative (Tang et al., 2006; WSD, 2009; Leung et al., 2012). When saline water is the only water source used for toilet flushing, already up to 30% of the fresh water use can be replaced by saline water (Lazarova et al., 2003). Although the use of saline water seems to be promising from a water consumption perspective, the salts originating from the saline water will eventually end up in the wastewater and reach the wastewater treatment plants, where they may affect the biological wastewater treatment processes.

Besides the direct use of saline water in urban environments, saline wastewater is generated by some industries such as the food production and processing (dairy industry, fish processing, pickled vegetables and meat canning) and tanneries (Gonzalez et al., 1983; Orhon et al., 1999; Fahim et al., 2000; Lefebvre et al., 2006). Furthermore, the intrusion of saline water, such as brackish ground water or seawater during high tides, into the sewerage systems leads to an increase of the salinity levels in the wastewaters transported by these systems.

To protect the surface water bodies against eutrophication, the removal of nitrogen (N) and phosphorus (P) from wastewaters needs to be implemented. Despite that several studies have focussed on the treatment of saline wastewater for COD removal and conventional N-removal via nitrification and denitrification, only few studies included effects of salinity on the enhanced biological phosphorus removal (EBPR). Moreover, the findings of those studies are inconsistent and non conclusive because other factors might have affected the EBPR as well. In some cases, the operating conditions were not optimal for the enrichment of PAO (e.g. lack of appropriate carbon source, high temperature and presence of NO_3 in the anaerobic phase). Based on the observations, presumably PAO were not the dominant microorganisms in those studies (Uygur and Kargi, 2004; Kargi and Uygur, 2005; Uygur, 2006; Panswad and Anan, 1999; Hong et al., 2007; Wu et al., 2008; Intrasungkha et al., 1999). In other studies, the salinity effects on COD and N removal processes may have caused negative side effects on PAO, such as the accumulation of nitrite (NO_2^--N) a known inhibitor of PAO (Pijuan et al., 2010) due to incomplete nitrification (Intrasungkha et al., 1999; Wu et al., 2008 and Cui et al., 2009) and lack of VFA in the anaerobic stage due to inhibited VFA production (Hong et al., 2007; Wu et al., 2008). In a long term study with complementary experiments, Pronk et al. (2014) demonstrated that elevated salinity concentrations (21g/L NaCl) led to a cascade inhibition effect, where the deterioration of nitrite oxidation resulted in nitrite accumulation which in turn severely affected EBPR. Furthermore, relevant stoichiometric and kinetic information needed to assess the impact of salinity on the metabolism of PAO (like the specific rates) are not reported in most past studies (Panswad and Anan, 1999; Intrasugkha et al., 1999; Uygur and Kargi, 2004; Kargi and Uygur, 2005; Kargi and Uygur, 2006; Hong et al., 2007; Cui et al., 2009).

In a recent study the impact of salinity on the anaerobic metabolism of PAO was assessed using an enriched PAO culture (Welles et al., 2014). In that study, it was demonstrated that the anaerobic metabolism of PAO is significantly affected by elevated salinity levels after a short-term exposure. The maximum acetate uptake rate and phosphate release rate of PAO decreased by 71% and 81%, respectively, when the salinity increased to 1% salinity. The anaerobic maintenance requirements and stoichiometry of the metabolic conversions were affected by salinity as well. However, a similar study on the aerobic metabolism of PAO is not yet available.

To get a better understanding of the salinity effects on the EBPR process, the impact of salinity on the aerobic metabolism (stoichiometry and kinetics) of an enriched PAO culture was investigated. Thus, aerobic batch tests were executed to assess how salinity affects the aerobic physiology of PAO regarding their maintenance requirements, initial aerobic kinetic rates and the stoichiometry of the conversions. This can contribute to assess the stability and feasibility to perform EBPR when treating saline wastewater in sewage plants not exposed regularly to high salinity concentrations.

7.2 Material and Methods

7.2.1 Enrichment of the PAO culture

7.2.1.1 Operation of SBR

This study is a continuation of a previous study about the salinity effect on the anaerobic metabolism of PAO II (Welles et al., 2014). Therefore detailed procedures for the enrichment of the PAO culture and the operation of the SBR are described in Welles et al. (2014). A PAO culture was enriched in a double-jacketed 2.5 L laboratory sequencing batch reactor (SBR). The reactor was operated and controlled automatically in a sequential mode (SBR). Activated sludge from a municipal wastewater treatment plant with a 5-stage Bardenpho configuration (Hoek van Holland, The Netherlands) was used as inoculum for the enrichment of the PAO culture. The SBR was operated in cycles of 6 hours (2.25h anaerobic, 2.25h aerobic and 1.5h settling phase). The pH was maintained at 7.2 ± 0.05, the Temperature was controlled in the reactor at 20 ± 1 °C. The SBR was controlled at a hydraulic retention time (HRT) of 12 h and at a sludge retention time (SRT) of 8 days.

The concentrated medium was prepared according to the procedures described in Welles et al. (2014). After dilution with demineralized water, the influent contained per litre: 860 mg NaAc·3H$_2$O (12.6 C-mmol/L, equivalent to 405 mg COD/L), 107 mg NH$_4$Cl (2 N-mmol/L), 89 mg NaH$_2$PO$_4$.H$_2$O (0.65 P-mmol/L, 20 mg PO$_4^{3-}$-P/L), 90 mg MgSO$_4$.7H$_2$O, 140 mg CaCl$_2$.2H$_2$O, 360 mg KCl , 2 mg of allyl-N-thiourea (ATU) and 0.3 mL trace element solution (Smolders et al.,1994a).

7.2.1.2 Monitoring of SBR

The performance of the SBR was regularly monitored by measuring orthophosphate (PO$_4$ -P), acetate (Ac-C), mixed liquor suspended solids (MLSS) and mixed liquor volatile suspended solids (MLVSS). Pseudo steady-state conditions in the reactor were confirmed based on a regular observation of the aforementioned parameters as well as pH and DO profiles. A cycle measurement was carried out to determine the biomass activity when the SBR reached steady-state conditions. In the cycle measurements, oxygen uptake rate (OUR), poly-hydroxyalkanoate (PHA) and glycogen were also measured. The degree of enrichment of the PAO culture was estimated via Fluorescence in situ Hybridization (FISH) analyses.

7.2.2.Aerobic batch tests

The kinetic rates and stoichiometry of the aerobic conversions at 0.02% salinity (the first salinity level) were determined in the SBR. The effect of higher salinity concentrations were assessed in batch experiments. After the biomass activity reached pseudo steady-state conditions in the SBR, aerobic batch experiments at different salt concentrations were performed in two double-jacketed laboratory reactors with a maximal operating volume of 0.5 L. In order to conduct aerobic short-term salinity tests, a defined volume of enriched PAO sludge was withdrawn from the SBR at the end of the anaerobic phase and transferred to the 0.5 L batch reactor. During the transfer of sludge, nitrogen gas was sparged to keep the anaerobic conditions. After each sludge transfer, the wastage of sludge in the parent SBR was adjusted to compensate for the sludge withdrawn to execute the batch tests and keep a stable SRT. Batch tests were performed at controlled temperature and pH (20 ± 0.5 °C and 7.0 ± 0.1, respectively). The pH was automatically controlled by dosing of 0.1 M HCl and 0.1 M NaOH. During the batch experiments, the sludge was constantly stirred at 300 rpm.

The batch tests carried out for the determination of the aerobic kinetic rates of PAO were executed after the addition of the identical synthetic media used for their cultivation in the parent SBR but without the presence of sodium acetate. In addition, NaCl was dosed at a certain defined concentration depending upon the salinity levels of interest. To suppress any potential foaming formation, a drop of antifoam (1% solution) was also added. The salinity concentrations studied (0.02, 0.19, 0.27, 0.35, 0.69, 1.02, 2.02 and 3.02 % (W/V) NaCl) were chosen following a step-wise approach, with the aim of minimizing the number of concentrations tested that can cover the full inhibition range from 0 to 100% inhibition. All batch tests were carried out for 16h, except for the tests at 0.35% and 0.69% salinity, which lasted only 3 h, due to system limitations. During the first 3 hours of each test (2.5 hour for the 0.02% salinity test performed in the SBR), MLVSS, MLSS, glycogen, PO$_4$ and NH$_4$ samples were taken and the O$_2$ consumption rate was determined throughout the test.

For the determination of the O$_2$ consumption rates in time, a 10 mL double jacketed respirometer, containing an oxygen sensor was connected to the batch reactor (Smolders et al., 1994; Brdjanovic et al., 1997). Sludge was recirculated for 1 min from the batch through the respirometer every 6 minutes in the beginning of the tests and every 20 minutes in the final part of the tests. Thus, the dissolved oxygen concentrations in the respirometer were measured for 5 or 19 minutes, depending on the recirculation frequency. For the determination of the maintenance oxygen consumption at 0.02% salinity, one additional test was conducted in the batch reactor.

7.2.3 Determination PAO maintenance coefficient

The PAO aerobic maintenance energy production rate ($m^o_{ATP,PAO_{O2}}$) of PAO was determined following a procedure described elsewhere (Brdjanovic et al., 1997; Lopez-Vazquez et al., 2007a). In each test (except those at 0.35% and 0.69% salinity) the qO$_2$ was followed during 16 hours. In most of the tests, the oxygen consumption rate became stable already after two hours. The oxygen consumption rate after 15 hours was taken as the oxygen consumption rate for maintenance energy production. At elevated salinity concentrations, PO$_4$ was released instead of taken up. This P-release was considered as an additional supply of energy. The maintenance energy production rates by poly-P consumption ($m^o_{ATP,PAO_{poly}-P}$) were determined by linear regression of the P-release profiles as described by Smolders et al. (1994a).

7.2.4 PAO Kinetics

The oxygen uptake rate (qO$_2$) was considered as the most relevant parameter because it represents all metabolic activities, i.e. poly-P formation, growth, glycogen formation and maintenance. Therefore the oxygen uptake rate was determined throughout the tests. To clarify the effect of salinity on the different metabolic processes, the initial NH$_4$-uptake rate (representing growth) and PO$_4$-uptake rate (representing poly-P formation) were the kinetic processes of interest in this study. In addition, the specific net-glycogen production was determined from the different batch tests. The specific kinetic rates were calculated considering the corresponding NH$_4$ or PO$_4$ and O$_2$ consumption profiles and the active biomass concentration from each aerobic batch test and the specific net-glycogen production was calculated with the net glycogen conversion and the active biomass concentration from each aerobic batch test. To obtain the oxygen consumption rate associated with poly-P formation, growth and glycogen formation, the corrected oxygen uptake rate was determined by subtracting the maintenance oxygen consumption rate from the total oxygen consumption rate. Similarly, the corrected phosphate uptake rate was determined by subtracting the P-release due to maintenance purposes from the negative P-uptake rates.

7.2.5 PAO Stoichiometry

To make a comparison to stoichiometric values of PAO and GAO reported in literature and to obtain a better understanding which processes were most severely inhibited by salinity, the following parameters were of interest: 1) total P-uptake/total oxygen uptake 2) corrected P-uptake/corrected oxygen consumption, 3) N-uptake/total oxygen consumption, 4) N-uptake/corrected oxygen consumption, 5) glycogen production/total oxygen consumption, 6) glycogen production/corrected oxygen consumption, 7) total P-uptake/glycogen production, 8) N-uptake/glycogen production, 9) N-uptake/total P-uptake.

7.2.6 Active Biomass

The active biomass concentration was determined following the procedures described in section 2.2.4 (chapter 2).

7.2.7 Analysis

PO_4^{3-}-P was determined by the ascorbic acid method. NH_4-N was analysed by the spectrophotometric method. Analyses, including MLSS and MLVSS determination, were performed in accordance with Standard Methods (A.P.H.A., 1995). Dissolved oxygen was measured using a WTW multiline (Multi 3410) portable oxygen meter. Glycogen analysis was executed according to the method described by Smolders et al. (1994) but with a digestion phase extended to 5 h. The PHA content (as PHB and PHV) of freeze dried biomass was also determined according to the method described by Smolders et al. (1994a).

To determine the microbial population distribution of the enriched PAO culture, Fluorescence *in situ* Hybridization (FISH) microscopy and quantification was performed according to the procedures described in Welles et al. (2014).

7.2.8 Parameter fitting and evaluation of models

With the aim to obtain a better understanding of the salinity effects on the aerobic metabolism and to propose a model for practical applications, a structured metabolic model was developed (Appendix 7 A-B), describing the salinity effects on the different metabolic processes. The aim of this model was to describe the salinity effect on the initial rates (PO_4-, NH_4- and O_2-uptake rates), the maintenance coefficients and the interrelationship of the different metabolic processes at increased salinity concentrations (Equation 1 - 4), assuming that the stoichiometric coefficients like the maximum yield coefficients (biomass, poly-P and glycogen yield on PHA), and the mechanistic relation between the storage polymer content and the kinetic rates as such are not affected by salinity.

7.2.8.1 PAO Aerobic Maintenance Coefficient

$$m_{ATP,PAO_{O2}}^{o}(S) = (m_{ATP,1}^{0,0} + a_1 \cdot S) \cdot \frac{1}{1 + e^{(bi_1 \cdot (S - Si_1))}} \qquad \text{(Equation 1)}$$

With,

$m_{ATP,PAO_{O2}}^{o}(S)$: PAO aerobic maintenance energy generated by O_2 consumption at different salinity concentrations

S: Salinity concentration

And fitted parameters,

$m_{ATP,1}^{0,0}$: O_2 maintenance coefficient at 0% salinity concentration

a_1: O_2 linear proportional increase in maintenance requirements per increase in salinity

bi_1: Impact factor, describing the magnitude of the inhibition effect on O_2 maintenance

Si_1: Inhibition factor, determining the salinity range at which the inhibition starts

7.2.8.2 PAO Aerobic kinetic rates

$$q_{O2,PAO\_cor}^{o}\ (S) = q_{O2,PAO\_P,N,Gly}^{0,0} * \frac{1}{1+e^{(bi_2.(S-Si_2))}}\ + q_{O2,PAO\_res}^{0,0} \qquad \text{(Equation 2)}$$

$$q_{PO4,PAO\_cor}^{o}(S) = q_{PO4,PAO\_cor}^{0,0} * \frac{1}{1+e^{(bi_3.(S-Si_3))}} \qquad \text{(Equation 3)}$$

$$q_{NH4,PAO}^{o}(S) = q_{NH4,PAO}^{0,0} * \frac{1}{1+e^{(bi_4.(S-Si_4))}} \qquad \text{(Equation 4)}$$

With,

$q_{O2,PAO\_cor}^{o}(S)$: PAO O_2 uptake rate at different salinity, corrected for maintenance O2 consumption

$q_{PO4,PAO\_cor}^{o}(S)$: PAO PO_4 uptake rate at different salinity, corrected for maintenance P-release

$q_{NH4,PAO}^{o}(S)$: PAO NH_4 uptake rate at different salinity

S: Salinity concentration

And fitted parameters,

$q_{O2,PAO\_P,N,Gly}^{0,0}$: PAO O_2 uptake rate at 0% salinity for PO_4, NH_4 and glycogen

bi_2: Impact factor, describing the magnitude of the inhibition effect on the O_2 uptake rate

Si_2: Salinity concentration at which 50% inhibition occurs of the O_2 uptake rate, which is associated with poly-P formation, growth and glycogen formation

$q_{O2,PAO\_res}^{0,0}$: PAO residual O_2 uptake rate

$q_{PO4,PAO\_cor}^{0,0}$: PAO PO_4 uptake rate at 0% salinity corrected for PO_4 maintenance

bi_3: Impact factor, describing the magnitude of the inhibition effect on the PO_4 uptake rate

Si_3: Salinity concentration at which 50% inhibition of the PO_4 uptake rate occurs

$q_{NH4,PAO}^{0,0}$: PAO NH_4 uptake rate at 0% salinity

bi_4: Impact factor, describing the magnitude of the inhibition effect on the NH_4 uptake rate

Si_4: Salinity concentration at which 50% inhibition of the NH_4 uptake rate occurs

The method of least squares was used to fit the model parameters with the experimental values of the kinetic rates at different salinity concentrations. The model simulation was evaluated using the ordinary least squares regression model (Mesple et al., 1996).

7.3 Results

7.3.1 Enrichment of PAO

The PAO SBR was operated for more than 660 days and pseudo steady-state conditions were confirmed before the experiments started. The conversions during a characteristic cycle are shown in Figure 7.1a. At the end of the aerobic phase, the MLSS and MLVSS concentrations were 3547 and 2447 mg/L, respectively, and the active biomass concentration was estimated of about 63 C-mol/L. The ash content of the sludge was 0.31gAsh/gTSS, which is characteristic of enriched PAO systems, indicating a high P-content. P-release during the anaerobic phase was 2.15 mmol P/L, and during the aerobic phase complete P removal was observed. P/HAc ratio was 0.34 P-mol/C-mol, which is similar to the ratio observed in a study with a highly enriched PAO II culture (Welles et al., 2014). In the anaerobic phase, the specific HAc-uptake rate and PO_4-release rate were 0.22 C-mol/(C-mol.h) and 0.082 P-mol/(C-mol.h), respectively. In the aerobic phase, the specific PO_4- and NH_4-uptake rates were 0.083 (P-mol/(C-mol. h)) and 0.0038 (N-mol/(C-mol. h)), respectively. In the FISH analyses (Figure 7.1b) the PAOmix probes (in red) showed a similar coverage as the EUB probes (in blue), indicating that the SBR was highly dominated by PAO (purple, the overlay of red and blue). The PAO fraction, determined by quantitative FISH was 99% (s.d. 3%) of the total microbial community. An additional FISH analysis showed more specifically that the reactor was dominated by *Candidatus* Accumulibacter phosphatis Type II (data not shown). GAO and PAO Type I were not observed.

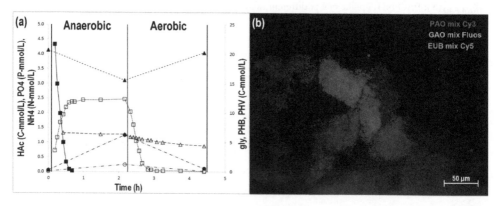

Figure 7.1 Enrichment of PAO culture: (a) Profiles observed during a cycle under pseudo steady-state conditions in the SBR: acetate (■), PO₄ (□), NH₄ (Δ) , PHB (●), PHV (○), and glycogen (▲) concentrations; (b) FISH image obtained from the PAO SBR: PAO mix (purple), GAO mix (cyan green), and EUB (blue)

7.3.2 Aerobic maintenance coefficient

In all batch tests the average oxygen uptake rate was determined in the 2nd, 3rd and the 16th hour (data not shown), except for the tests performed at 0.33% and 0.67% salinity due to problems in the batch test. At 0.02, 0.18 and 0.27% salinity, the oxygen uptake rate became stable after several hours, indicating that the metabolic processes of growth, poly-P formation and glycogen formation were finished and that this oxygen uptake rate was likely due to maintenance. In the tests with salinity concentrations above 0.33% salinity, PO$_4$ uptake (representing poly-P formation), NH$_4$ uptake (representing growth) and glycogen formation were not observed. Consequently, the oxygen consumption rate observed after 1 hour was mainly caused by the maintenance activity. This was confirmed by the rather similar average oxygen uptake rates after the 2nd, 3rd and 16th hour. Interestingly, the maintenance oxygen consumption rate in the test at 0.02% salinity was almost double compared to the oxygen consumption at 0.18% salinity (Figure 7.2).

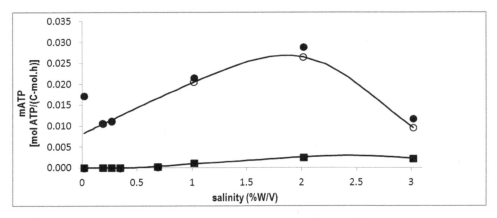

Figure 7.2 Effect of salinity on the aerobic maintenance coefficients of PAO: $m^o_{ATP,PAO_{poly}-P}$ (■), $m^o_{ATP,PAO_{O2}}$ (○), and $m^o_{ATP,PAO_{total}}$ (●), the solid lines show the simulation of the maintenance coefficient using the expression proposed in this study (Appendix 7A, Equation 1)

The average oxygen uptake rates in the 16th hour were taken as the oxygen consumption rates for the determination of the aerobic ATP maintenance coefficient ($m^o_{ATP,PAO_{O2}}$) (Figure 7.2). Above 0.33% salinity aerobic P-release took place instead of aerobic P-uptake. Thus, the maintenance energy produced by the observed P-release ($m^o_{ATP,PAO_{poly}-P}$) was determined as well (Figure 7.2). Consequently, the total ATP generated to cover the aerobic maintenance requirements ($m^o_{ATP,PAO_{total}}$) was estimated as the sum of the ATP generated by the aerobic P-release and by the oxygen consumption rate. Figure 7.2 shows that the total aerobic maintenance energy production of PAO increased and was estimated to be about 0.025 mol ATP/(C-mol.h) at 2% salinity but decreased rapidly to around 0.012 mol ATP/(C-mol.h) at 3% salinity. The contribution of maintenance energy production by poly-P consumption to the total maintenance energy production seems to be negligible. Therefore, the effects of salinity on the aerobic maintenance coefficient, reflecting only the oxygen consumption of PAO were described successfully by Equation 1 (Appendix 7A) using the parameters shown in Table 7.1. The ATP maintenance coefficient determined at 0.02% salinity from the oxygen uptake rates (0.017 ATP-mol/(C-mol.h) was not included in the model parameter estimation because it was unclear whether the observed oxygen consumption rate was solely for maintenance purposes or that it was partly due to the occurrence of other metabolic processes. The increase of the maintenance energy production by oxygen consumption represented by a_1, shows that the energy requirments at 0.02% salinity ($m^{o,0}_{ATP,1}$) increase by a factor 3 when the salinity concentration increases up to 2%

Table 7.1 Fitted parameters of the equations describing the PAO maintenance energy production by O2 consumption at different salinities.

Kinetic	Parameter	Unit	Value
Maintenance rate	$m_{ATP,1}^{0,0}$	[ATP-mol/(C-mol.h)]	0.0081
	a_1	[ATP-mol/(C-mol.h.(%W/V))]	0.013
	bi_1	[1/%(W/V)]	2.67
	Si_1	(% W/V)	2.52
Evaluation of total maintenance rates simulation	RSQ		0.9995
	Slope	N.A.	0.9995
	intercept		0.0000

N.A.: Not Applicable

7.3.3 Impact of salinity on the aerobic kinetic rates

The oxygen uptake rate was considered to be the most important parameter because it is a reflection of all the aerobic metabolic processes, i.e. maintenance, poly-P recovery, glycogen recovery and growth. Figure 7.3 shows the specific total oxygen uptake rate ($q_{O2,PAO}^o$) observed in the batch tests at different salinity concentrations and the oxygen uptake rate corrected for oxygen consumption due to maintenance ($q_{O2,PAO_{cor}}^o$). At 0.02% salinity, the oxygen profiles showed the highest oxygen uptake rates in the beginning of the tests, likely associated with poly-P formation, growth, and glycogen formation. After half an hour, the oxygen uptake rates rapidly decreased as all orthophosphate is taken up and probably PHA, the carbon and energy source under aerobic conditions, gets depleted. However, at 0.18% salinity, the biomass activity starts to be inhibited. $q_{O2,PAO}^o$ and $q_{O2,PAO_{cor}}^o$ are lower and, consequently, due to the potentially slower PHA utilization and P-removal, a continuous linear decrease in the respiration rates is observed for almost 1.25-1.5 h before it levels off. Above 0.18% salinity, the inhibition was more drastic, much lower oxygen uptake rates were measured (despite that enough PHA was presumably available) that stopped after 30 min at the different salinities tested (up to 3.02%).

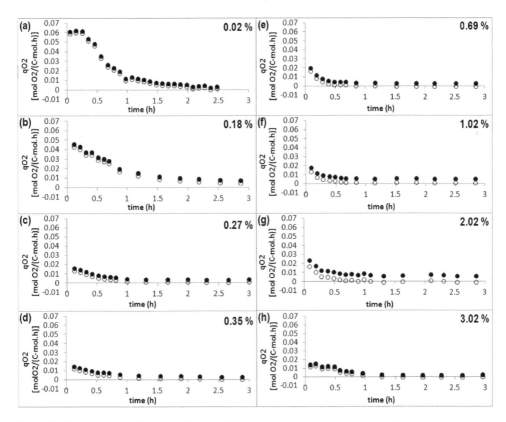

Figure 7.3 Total oxygen uptake rates ($q^o_{O2,PAO}$ (●)) and oxygen uptake rates corrected for maintenance oxygen consumption (q^o_{O2,PAO_cor} (○)) as a function of time at different salinity concentrations

Figure 7.4 displays the maximum specific rates of (a) corrected oxygen uptake (corrected for oxygen consumption due to maintenance) (b) corrected phosphate uptake (corrected for maintenance P-release which was observed at higher salinity concentrations), (c) ammonium uptake, and (d) specific net-glycogen production. The maximum oxygen uptake rates (corrected for maintenance) decreased around 25% when the salinity concentration increased from 0.02 to 0.18%, but decreased up to 80% at 0.35% salinity. Above 0.35% salinity, the corrected oxygen uptake rate remained constant at around 0.017 mol O_2/(C-mol.h) (Figure 7.4a), indicating that some processes other than maintenance were still taking place. Remarkably, the profiles of the initial rates or net conversions of the other known processes, e.g. growth, poly-P formation and glycogen formation, suggest that these processes were fully inhibited above 0.35% salinity. Interestingly, at 0.18% salinity, the corrected PO_4 uptake rate (corrected for maintenance P-release) decreased by 46% (Figure 7.4b), and above 0.35% salinity, PO_4 was released (Figure 7.2). The NH_4 uptake rate (an indirect parameter related to biomass growth) was more sensitive to salinity. At 0.18% it was 63% inhibited and just at 0.18% salinity no NH_4 uptake was longer observed (Figure 7.4c). Due to unknown reasons, the NH_4 data at 0.25% salinity suggests that certain NH_4 release took place, but considering the trend of the NH4 uptake rates, the data at 0.25% salinity seems to be an outlier. Regarding glycogen formation, the specific net-glycogen production during the tests was determined (Figure 7.4d). The net glycogen production profile shows that above 0.18% salinity the specific net-glycogen formation decreased drastically. The specific net-glycogen consumptions observed at 0.25% salinity and 0.69% salinity, suggest that glycogen was consumed instead of being produced, but these observations are not supported by the trend of the specific net-glycogen production. Above 0.69% salinity no significant net-glycogen formation occurred. Actually, all the aerobic processes studied (PO$_4$-uptake, NH$_4$-uptake, and glycogen production) were fully inhibited above 0.35% salinity.

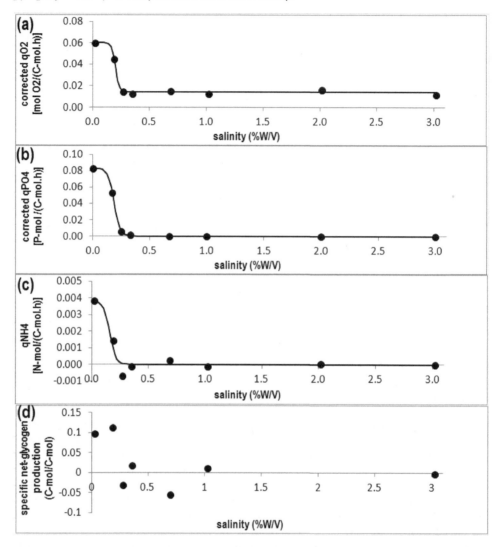

Figure 7.4 Salinity effects on the aerobic metabolism of PAO on: (a) oxygen uptake rate corrected for maintenance oxygen consumption, (b) PO$_4$ uptake rate corrected for maintenance PO$_4$ release, (c) NH4 uptake rate and (d) specific net-glycogen production. Continuous bold lines show the results of model simulations

The corrected oxygen uptake rates, corrected PO$_4$ uptake rates and NH$_4$ uptake rates were successfully described by equations 2, 3 and 4 (Appendix 7B) using the fitted parameters shown in Table 7.2. The Si values, indicating the concentration at which 50% inhibition occurs, show that the NH$_4$-uptake rate was the most sensitive (with a Si of 0.17%), followed by the PO$_4$-uptake rate and the oxygen uptake rate associated with poly-P formation, growth and glycogen formation) (both with a Si of 0.20%). The residual oxygen consumption rate $q_{PO4,PAO\_res}^{0,0}$ was considered to be constant.

Table 7.2 Fitted parameters of the equations describing specific aerobic O₂, PO₄ and NH₄ uptake rates of PAO at different salinities.

Kinetic	Parameter	Unit	Value
O₂ uptake rates	$q_{O2,PAO\_P,N,Gly}^{o,0}$	[mol O₂/(C-mol.h)]	0.046
	bi_2	[1/%(W/V)]	61
	Si_2	(% W/V)	0.20
	$q_{PO4,PAO\_res}^{o,0}$	[mol O₂/(C-mol.h)]	0.014
Evaluation of oxygen uptake rate	RSQ		0.9935
	Slope	NA	0.9935
	Intercept		0.0002
PO₄ uptake rates	$q_{PO4,PAO\_cor}^{o,0}$	[P-mol/(C-mol.h)]	0.083
	bi_3	[1/%(W/V)]	36
	Si_3	(% W/V)	0.20
Evaluation of PO₄ uptake rates simulation	RSQ		0.9995
	Slope	N.A.	1.0076
	Intercept		-0.0006
NH4 uptake rate	$q_{NH4,PAO}^{o,0}$	[N-mol/(C-mol.h)]	0.0038
	bi_4	[1/%(W/V)]	36
	Si_4	(%W/V)	0.17
Evaluation of NH4 uptake rates simulation	RSQ		0.9937
	Slope	N.A.	0.9928
	Intercept		0.0000

N.A.: Not Applicable

7.3.4 Impact of salinity on the aerobic stoichiometry

The stoichiometry of the aerobic conversions is shown in Figure 7.5. The stoichiometric values were only shown in the salinity range 0.02% to 0.35% salinity, where the net-conversions were still detectable. When the denominator of the ratios dropped below zero, the values are not presented. From salinity concentrations slightly lower than 0.20%, the total PO₄/corrected O₂ and NH₄/corrected O₂ decreased (Figures 7.5a and 7.5b), while the glycogen/corrected O₂ was less affected (Figure 7.5c). At 0.25% salinity the observed gly/corrected O₂ and NH₄/corrected O₂ stoichiometric values were negative because the net-glycogen production and NH₄-uptake values determined at this salinity concentration were negative due to unknown reasons. The total PO₄/gly and NH₄/gly (only determined at 0 and 0.18% salinity) stoichiometry profiles (Figure 7.5d) showed a decrease at elevated salinity concentrations and the NH₄/PO₄ (Figure 7.5d) decreased as well.

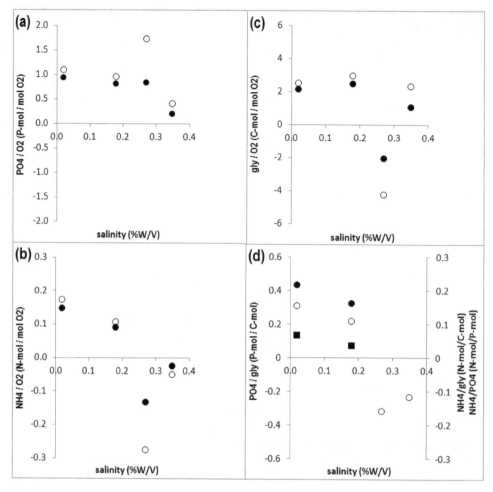

Figure 7.5 Stoichiometry of aerobic conversions of PAO at different salinity concentrations: (a) total PO_4/total O_2 (●) and total PO_4/O_2 corrected for maintenance (○); (b) NH_4/total O_2 (●) and NH_4/O_2 corrected for maintenance (○); (c) gly/total O_2 (●) and gly/O_2 corrected for maintenance (○), and; (d) total PO_4/gly (●), total NH_4/PO_4 (○) and NH_4/gly (■)

7.4 Discussion

7.4.1 Enrichment of PAO

The PAO culture enriched in this study, *Candidatus* Accumulibacter phosphatis Type II, was the same culture as the one used to study the short-term salinity effects on PAO anaerobic metabolism (Welles et al., 2014). In Table 7.3, the aerobic kinetic rates and stoichiometric values obtained in this study were compared to those from previous studies (Smolders et al., 1994; Brdjanovic et al., 1997). As observed, the experimentally determined $m^o_{ATP,PAO_{O2}}$ was in the range of the $m^o_{ATP,PAO_{O2}}$ values obtained for PAO by Smolders et al. (1994) and Brdjanovic et al. (1997). Remarkably, the ATP aerobic maintenance coefficient determined in this study from the oxygen uptake rate (0.017 mol ATP/(C-mol.h)) was 4.4 times higher than the ATP anaerobic maintenance coefficient (0.0039 mol ATP/(C-mol.h) estimated in a previous study with the same sludge (Welles et al., 2014). Similarly, Smolders et al. (1994[b]) observed that the ATP aerobic maintenance coefficient (0.019 mol ATP/ (C-mol.h) was 7.6 times higher than the ATP anaerobic maintenance coefficient (0.0025 mol ATP/(C-mol.h)) (Smolders et al., 1994[a]). However, the ATP aerobic maintenance coefficient determined at 0.02% salinity from the ATP aerobic maintenance coefficient trend by model parameter estimation (0.0083 mol ATP/(C-mol.h)) was only 2.1 times bigger than the ATP anaerobic maintenance coefficient (0.0039 mol ATP/(C-mol.h). This suggests that the apparently stable oxygen consumption rate, often used to determine the oxygen consumption for maintenance, might still cover additional metabolic processes such as growth. Oehmen et al. (2007) demonstrated in an experiment with an extended aerobic phase with measurements of all carbon conversions, i.e., PHA, glycogen and CO_2, that indeed some growth and glycogen consumption continue to take place. Once salinity increases and the metabolic processes, like growth and poly-P formation are suppressed or completely inhibited in this study, then a stable oxygen profile will be a true reflection of the maintenance processes.

Regarding the kinetic rates, the oxygen and phosphate uptake rates were about two times higher than the rates obtained by Brdjanovic et al. (1997), while the ammonium uptake rate was similar. The stoichiometric ratios of the aerobic conversions determined in this study are comparable to those observed by Smolders et al. (1994), except for the specific net-glycogen recovery that almost doubled the specific net-glycogen production observed by Smolders et al. (1994). The different kinetic and stochiometric values reported might be caused by the possibility that different PAO clades were enriched in each study. Intrinsic differences in the physiological and morphological properties of PAO clades observed in previous studies (Flowers et al., 2009; Slater et al., 2010 and Carvalho et al., 2007) can support this hypothesis. In this study a highly enriched PAO clade II culture was obtained, but unfortunately the clade differentiation is only recently possible and comparison to older literature (Smolders et al. 1994 and Brdjanovic et al., 1997) is therefore hampered.

7.4.2 Effects of salinity on the aerobic maintenance coefficient

The increase in the maintenance oxygen uptake rate from 0.16 to 2% salinity indicates that salinity increases the maintenance energy requirements. A similar phenomenon was observed on the anaerobic maintenance energy requirements of PAO (Welles et al., 2014). As explained elsewhere (Welles et al., 2014), the leakage of salt ions through the cell membrane forces the organisms to expel the ions at the expense of additional energy requirements for maintenance (Castle et al. 1986). Similar to the increased anaerobic maintenance requirements but with a different sensitivity, the maintenance oxygen uptake rate decreased above 2 % salinity. This was previously explained by the inhibition of the maintenance activity, but such an event would eventually lead to a progressive inhibition followed by the complete inactivation of the cells. Since the oxygen consumption rate at 3% salinity was stable during 15 hours, there was no indication of any progressive inhibition. Alternatively, the maintenance energy requirements at high salinity concentrations could be lower than the energy requirements under medium salinity levels as a consequence of shrinking cells volumes and more compact cell membranes due to the osmotic pressure. Interestingly, above 0.35% salinity, PO_4 was released instead of being taken up, suggesting that the cells needed additional energy from PO_4 release for detoxification of the cells. Furthermore, at 0.35% and 0.69% salinity glycogen consumption was measured during the first three hours of the tests, which could be explained as a metabolic shift from PHA oxidation to glycogen oxidation or to conversion of glycogen into PHA to compensate a potential inhibition of the respiratory pathways or a depletion of certain storage polymers (Lopez et al., 2006; Lu et al., 2007; Vargas et al., 2013) due to higher maintenance energy requirements. However, the trend of glycogen consumption versus salinity concentrations was not consistent and PHA concentrations were not measured to support these observations. Nevertheless, it was confirmed that glycogen did not deplete (data not shown) in any of the tests and glycogen depletion can therefore be discarded as a factor triggering a potential metabolic shift from glycogen consumption to poly-P consumption. As the specific net-oxygen consumption determined in each test (data not shown) drastically decreased at elevated salinity concentrations, it is also not likely that PHA got depleted, triggering   a   shift   from   PHA   to   glycogen   oxidation.

Table 7.3 PAO kinetic and stoichiometric parameters observed in this study and reported in literature under similar operating conditions (pH 7 and 20oC).

Salinity	organism	SRT	HRT	$m^o_{ATP,PAO,total}$	Initial rates			specific net-gly production	total PO4 / total O2	NH4 / total O2	Gly / total O2	total PO4 / gly	NH4 / gly	total NH4 / PO4	Reference
					$q_{O2,total}$	$q_{PO4,total}$	q_{NH4}								
(%W/V)		days	hours	[mol ATP / (C-mol.h)]	[mol O2 / (C-mol.h)]	[P-mol / (C-mol.h)]	[N-mol / (C-mol.h)]	(C-mol/C-mol)	[P-mol / mol O2]	[N-mol / mol O2]	[C-mol / mol O2]	[P-mol / C-mol]	[N-mol / C-mol]	[P-mol / N-mol]	
0.017	PAO	8	12	0.019	-	-	-	N.A.	0.91[a]	0.14[a]	1.22[a]	0.75[a]	0.12[a]	0.15[a]	Smolders et al. (2004)
0.017	PAO	8	12	0.012	0.031	0.037	0.0030	0.060[a]	1.20	0.10	1.28	0.93[a]	0.078[a]	0.083[a]	Brdjanovic et al.(1997)
0.017[b]				0.017[b], 0.0083[b,c]	0.062[b]	0.083[b]	0.0038[b]	0.098[b]	0.94[b]	0.15[b]	2.2[b]	0.44[b]	0.068[b]	0.16[b]	
0.18				0.010	0.047	0.053	0.0014	0.11	0.82	0.092	2.5	0.33	0.037	0.11	
0.27				0.011	0.017	0.0065	-0.0007	-0.031	0.85	-0.13	-2.0	n.d.	n.d.	-0.16	
0.35	PAO	8	12	0.012*	0.016	0.0022	-0.0001	0.018	0.2	-0.023	1.1	n.d.	n.d.	-0.11	This study
0.69				0.017*	0.020	-0.00017	0.0002	-0.055	n.d.	n.d.	n.d.	n.d.	n.d.	n.d.	
1.0				0.022	0.018	-0.0011	-0.0001	0.011	n.d.	n.d.	n.d.	n.d.	n.d.	n.d.	
2.0				0.029	0.023	-0.0025	0	n.d.	n.d.	n.d.	n.d.	n.d.	n.d.	n.d.	
3.0				0.012	0.015	-0.0022	0	-0.0022	n.d.	n.d.	n.d.	n.d.	n.d.	n.d.	

n.d.: Not determined

N.A.: Not applicable

[a]: calculated

[b]: values obtained from SBR

[c]: estimated by model

7.4.3 Effects of salinity on the aerobic kinetic rates

The impact of salinity on PAO's aerobic activity was pronounced. Both the NH_4- and PO_4 uptake rates, reflecting growth and poly-P formation, respectively, were affected significantly. The corrected oxygen uptake rates were less affected. Although poly-P formation, growth and glycogen formation seemed to be fully inhibited above 0.35% salinity, an initial residual corrected oxygen uptake rate was still observed, which could not clearly be associated with any known metabolic process. Since the residual corrected oxygen uptake rate seemed to be rather insensitive to salinity, it cannot provide a direct explanation based on the initial stress response to salinity.

In comparison to the short-term salinity effects on the anaerobic metabolism of PAO, the inhibition of the aerobic kinetic rates, seems to follow a similar trend but with a higher sensitivity to salinity. The anaerobic HAc-uptake and P-release rates were 50% inhibited at salinity concentrations of 0.6% salinity, while 50% inhibition of the aerobic PO_4 and NH_4 uptake occurs just at 0.2 % salinity (Welles et al., 2014).

7.4.4 Effects of salinity on the aerobic stoichiometry

The NH_4/O_2 stoichiometry was affected mostly and the gly/O_2 stoichiometry affected the least, indicating that growth was the most sensitive process with respect to salinity, followed by poly-P recovery and glycogen recovery. This hypothesis was supported by the PO_4/gly, NH_4/gly and NH_4/PO_4 stoichiometry profiles, which showed a decrease when increasing the salinity concentrations. This preferential inhibition could be explained by the differences in energy demand for each process. In the study by Smolders et al. (1994), energy demands of 1.6 mol ATP/C-mol active biomass, 1.26 mol ATP/P-mol poly-P, and 0.83 mol ATP/C-mol gly, were observed for biomass growth, poly-P recovery and glycogen recovery, respectively. If PAO suffer from any energy limitation, it is logical that, according to their energy requirements in a decrease order, growth will be the most inhibited followed by poly-P formation and glycogen synthesis. Thus an inhibition of the aerobic energy generating pathway through O_2 consumption, as suggested by the observed P-release, would result in a lower ATP/ADP ratio and thereby trigger a selective pressure to specific pathways with lower energy requirements.

7.4.5 Modelling the aerobic maintenance and kinetic rates of PAO

The aerobic maintenance energy requirements and initial kinetic rates of PAO were successfully described by the proposed model equations. The model may be used to predict the (shock-load) salt effects up to 3% salinity on real wastewater treatment plants, but it needs to be validated using activated sludge from different full-scale EBPR systems.

7.4.6 Comparison with previous studies

This study was conducted on a highly enriched PAO II culture and therefore the effects of salinity could be assessed with a high degree of certainty on this organism. However, the disadvantage is that the overall effects on other PAO clades (probably more representative and abundant in activated sludge systems) may not be similar. Previous studies (Flowers et al., 2009; Slater et al., 2010 and Carvalho et al., 2007) have demonstrated that there are metabolic differences between the different PAO clades. In line with the metabolic differences observed in previous studies, different PAO clades may have different tolerance or response to salinity as well. Furthermore, the experiments conducted in this study were short-term experiments that do not allow the sludge to acclimatize to the higher salinity concentrations. The long-term exposure of PAO II to higher salinity concentrations could lead to their acclimatization and adaptation.

Although previous studies on biological nutrient removal with EBPR as an integral part of those systems were not conclusive because often additional factors negatively affected the PAO metabolism, one long-term study carried out in a MUCT configuration (Cui et al., 2009), showed that about 70% P-removal still occurred at 0.8% salinity (despite that certain operating factors were not favourable to PAO). Furthermore, a study performed by Bassin et al. (2011) using granular sludge, showed that the tolerance of PAO to salinity during short-term exposure was slightly higher than the tolerance observed in this study (i.e. the aerobic P-uptake rate was 70% inhibited at 0.5% salinity). But, after a long-term acclimatization, PAO became more tolerant and PO_4 uptake activity took place even up to 3% salinity. These observations supports the idea that PAO either have the ability to acclimatize to higher salinity or that specific PAO clades can have a higher tolerance to salinity than others.

7.5 Conclusions

The results of this study suggest that the sudden exposure of non-acclimatized activated sludge to salinity levels higher than 0.18% may have deleterious effects on biological phosphorus removal due to the direct inhibition of PAO by salinity. The maintenance profiles indicate that PAO are still able to stand and survive up to 2% salinity, suggesting that after a temporal salinity shock, PAO might be able to recover. These findings suggest that any saline discharge to an EBPR plant treating 'fresh' wastewater equivalent to more than 5% seawater (with 3.4% salinity) addition or 15% brackish water (with 1.2% salinity) by either seawater toilet flushing, industrial discharges or saline intrusion can cause serious upsets of the phosphorus removal process. To investigate if certain PAO species pose a higher tolerance to salinity and/or PAO can acclimatize, long-term studies need to be conducted on enriched PAO cultures under well defined environmental and operating conditions.

7.6 Acknowledgements

This research study was carried out as part of the SALINE project (http://www.salinesanitation.info) led by UNESCO-IHE Institute for Water Education and consortium partners KWR Watercycle Research Institute, Delft University of Technology, University of Cape Town, The Hong Kong University of Science and Technology, The Higher Polytechnic Institute "José Antonio Echeverría" (CUJAE) and Birzeit University. The SALINE project is financed by UNESCO-IHE internal research fund with a special generous contribution from Professor George Ekama from University of Cape Town. The authors would like to gratefully thank Professor George Ekama for his contribution. Thanks to UNESCO-IHE laboratory staff for all their support during the research project.

7.7 References

Amann R.I. (1995) In situ identification of microorganisms by whole cell hybridization with rRNA-targeted nucleic acid probes. In: Akkermans ADL, van Elsas JD, de Bruijn FJ (eds). Molecular microbial ecology manual. Kluwer Academic Publisher, London, pp 1-15

Bassin J.P., Pronk M., Muyzer, G. Kleerebezem R., Dezotti M., Van Loosdrecht M.C.M. (2011) Effect of Elevated Salt Concentrations on the aerobic granular sludge process: Linking Microbial Activity with Microbial Community Structure. Appl Environ Microbiol 77(22): 7942-7953

Brdjanovic D., van Loosdrecht M.C.M., Hooijmans C.M., Alaerts G.J., Heijnen J.J. (1997) Temperature Effects on Physiology of Biological Phosphrorus Removal. ASCE J Environ Eng-ASCE 123(2): 144-154.

Carvalho G., Lemos P.C., Oehmen A., Reis M.A.M. (2007) Denitrifying phosphorus removal: Linking the process performance with the microbial community structure. Water Res 41: 4383-4396

Castle A.M. Macnab R.M. Shulman R.G. (1986) Coupling between the Sodium and Proton Gradients in Respiring Escherichia coli Cells Measured by 23Na and 31P Nuclear Magnetic Resonance. J Biol Chem 261(17): 7797-7806

Crocetti G. R., Hugenholtz P., Bond P.L., Schuler A., Keller J., Jenkins D., Blackall L.L., (2000) Identification of polyphosphate-accumulating organisms and design of 16S rRNA-directed probes for their detection and quantitation. Appl. Environ Microbiol 66: 1175-1182

Cui Y., Peng C., Peng Y., Ye L. (2009) Effects of Salt on Microbial Populations and Treatment Performance in Purifying Saline Sewage Using the MUCT Process. Clean-Soil Air Water 37(8): 649-656

Fahim F.A., Fleita D.H. Ibrahim A.M., El-Dars F.M.S. (2000) Evaluation of some methods for fish canning wastewater treatment. Water Air Soil Pollut 127: 205-226

Flowers J.J., He S., Yilmaz S., Noguera D.R., McMahon K.D. (2009) Denitrification capabilities of two biological phosphorus removal sludges dominated by different 'Candidatus Accumulibacter' clades. Environ Microbiol Rep (2009) 1(6), 583-588.

Gonzalez J.F., Civit E.M., Lupin H.M. (1983) Composition of fish filleting wastewater. Water SA 9(2): 49-56

Hong C.C., Chan S.K., Shim H. (2007) Effect of chloride on biological nutrient removal from wastewater. J Appl Sci Environ Sanit 2(3): 85-92

Intrasungkha N., Keller J., Blackall L.L. (1999) Biological nutrient removal efficiency in treatment of saline wastewater. Water Sci Technol 39(6): 183-190

Kargi, F., A. Uygur, (2005) Improved Nutrient Removal from Saline Wastewater in an SBR by Halobacter Supplemented Activated Sludge. Envir. Eng. Sci. 22(2):170-176.

Lazarova V., Hills S., Birks R. (2003) Using recycled water for non-potable, urban uses: a review with particular reference to toilet flushing. Water Sci Technol 3(4), 69-77

Lefebvre O., Moletta R., (2006) Treatment of organic pollution in industrial saline wastewater: A literature review. Water Res 40: 3671-3682.

Leung R.W.K., Li D.C.H., Yu W.K., Chui H.K., Lee T.O., Van Loosdrecht M.C.M., Chen G.H. (2012) Integration of seawater and grey water reuse to maximize alternative water resource for coastal areas: The case of the Hong Kong International Airport. Water Sci Technol 65(3): 410-417

Lopez C., Pons M.N. Morgenroth E. (2006) Endogenous processes during long-term starvation in activated sludge performing enhanced biological phosphorus removal. Water Res 40(8): 1519-1530

Lu H., Keller J., Yuan Z. (2007) Endogenous metabolism of Candidatus Accumulibacter phosphatis under various starvation conditions. Water Res 41(20): 4646-4656

Mesple F., Troussellier M., Casellas C., Legendre P. (1996) Evaluation of simple statistical criteria to qualify a simulation. Ecol Model 88(1-3): 9-18

Oehmen A., Zeng R., Keller J., Yuan Z. (2007) Modeling the aerobic metabolism of polyphosphate-accumulating organisms enriched with propionate as a carbon source. Water Environ. Res. 79(13): 2477-2486

Orhon D., Tasli R., Sozen S. (1999) Experimental basis of activated sludge treatment for industrial wastewaters - the state of the art. Water Sci Technol 40(1): 1-11

Panswad T., Anan C. (1999) Impact of high chloride wastewater on an anaerobic/anoxic/aerobic process with and without inoculation of chloride acclimated seeds. Water Res 33(5): 1165-1172

Pijuan M., Ye L., Yuan Z. (2010) Free Nitrous acid inhibition on the aerobic metabolism of poly-phosphate accumulating organisms. Water res 44(20), 6063-6072

Pronk M., Bassin J.P., de Kreuk M.K., Kleerebezem R., van Loosdrecht M.C.M. (2013) Evaluating the main side effects of high salinity on aerobic granular sludge. Appl Mirobiol Biotechnol 98: 1339-1348

Slater F.R., Johnson C.R., Blackall L.L., Beiko R.G., Bond P.L. (2010) Monitoring associations between clade-level variation, overall community structure and ecosystem function in enhanced biological phosphorus removal (EBPR) systems using terminal-restriction fragment length polymorphism (T-RFLP). Water Res 44(17), 4908-4923

Smolders G.J.F., Van der Meij J., Van Loosdrecht M.C.M. and Heijnen J.J. (1994[a]) Model of the anaerobic metabolism of the biological phosphorus removal process: Stoichiometry and pH influence. Biotechnol Bioeng 43: 461-470

Smolders G.J.F., van der Meij M.C.M., van Loosdrecht M.C.M. Heijnen J.J. (1994[b]) Stoichiometric model of the aerobic metabolism of the biological phosphorus removal process. Biotechnol Bioeng 44(7): 837-848

Smolders G.J.F., van der Meij J., van Loosdrecht ,M.C.M. Heijnen J.J. (1995) A structured model for anaerobic and aerobic stoichiometry and kinetics of the biological phosphorus removal process. Biotechnol Bioeng 47(3): 227-287

Tang S.L., Yue D.P.T., Li X.Z. (2006) Comparison of engineering costs of raw freshwater, reclaimed water and seawater for toilet flushing in Hong Kong. Water Environ J 20(4): 240-247

Uygur A., Kargi F. (2004) Salt Inhibition on biological nutrient removal from saline wastewater in a sequencing batch reactor. Enzyme and Mic Tech 34: 313-318

Uygur A. (2006) Specific nutrient removal rates in saline wastewater treatment using sequencing batch reactor. Proc Bio 41(1): 61-66

7.8 Appendix 7A

The total maintenance energy of PAO (Equation 1a) is considered to be produced by PHA conversion with oxygen (Equation 1b) and at elevated salinity concentrations by both PHA conversion and poly-P cleavage (Equation 1c). The activation of the poly-P consumption at elevated salinity concentrations is described by a switch function as can be seen in Equation 1c, where Si_3 represents the salinity concentration at which the Poly-P consumption starts to occur. For both the PAO maintenance energy production by PHA conversion and maintenance energy production by poly-P cleavage, it was assumed that the maintenance activity is a product of the maintenance requirements and an empirical inhibition factor (Equation 1b and 1c). Due to the need to detoxify, the required maintenance activity increases with increasing salinity but above a certain salinity concentration, the maintenance activity becomes inhibited as well, leading to lower maintenance energy production rate. The parameter Si_2 represents the concentrations at which 50% inhibition of the required maintenance energy production rate occurs.

PAO Aerobic Maintenance Coefficient

$$m^o_{ATP,PAO_{total}}(S) = m^o_{ATP,PAO_{poly-P}}(S) + m^o_{ATP,PAO_{O2}}(S)$$ (Equation 1a)

$$m^o_{ATP,PAO_{O2}}(S) = (m^{o,0}_{ATP,1} + a_1.S).\frac{1}{1+e^{(bi_1.(S-Si_1))}}$$ (Equation 1b)

With,

$m^o_{ATP,PAO_{total}}(S)$ PAO total aerobic maintenance coefficient at different salinity concentrations

$m^o_{ATP,PAO_{O2}}(S)$: PAO aerobic maintenance energy generated by O_2 consumption at different salinity concentrations

$m^o_{ATP,PAO_{poly-P}}(S)$ PAO aerobic maintenance energy generated by poly-P consumption at different salinity concentrations$S$: Salinity concentration

And fitted parameters,

$m^{o,0}_{ATP,1}$: O_2 maintenance coefficient at 0% salinity concentration

a_1: O_2 linear proportional increase in maintenance requirements per increase in salinity

bi_1: Impact factor, describing the magnitude of the inhibition effect on O_2 maintenance

Si_1: Inhibition factor, determining in which salinity range the inhibition starts to occur of the O_2 maintenance

7.9 Appendix 7B

In the aerobic phase poly-P recovery, growth, glycogen recovery and maintenance occur. All these processes require PHA and oxygen consumption. Unfortunately, PHA data is lacking and the quality of the glycogen data was insufficient to determine glycogen production rates. Still the maintenance oxygen consumption rate, the initial PO_4-uptake rate (representing poly-P formation), NH_4 -uptake rate (representing growth) and O_2 uptake rates could be determined. It was expected that the corrected oxygen consumption rate (corrected for oxygen consumption due to maintenance) would become zero when the PO_4-uptake rate, the NH_4-uptake rate and net-glycogen production were fully inhibited. Assuming that the stoichiometric coefficients determined by Smolders et al. (1994) are unaffected, this data would allow then the estimation of the glycogen consumption rate. However, residual O_2 uptake rates were observed at elevated salinity concentrations, where the initial PO_4-uptake rate, the NH_4 -uptake rate and net-glycogen recovery seemed to be fully inhibited. This residual oxygen seemed to have no correlation with the increasing salinity concentration. Due to this residual O_2 uptake rate, it was impossible to estimate the glycogen production rate. Still it was possible to describe the inhibition of the initial PO_4-uptake rate (Equation 3) and NH_4 -uptake rate (Equation 4) with an inhibition equation, which was also used for the salinity effects on the anaerobic PO_4 release rate in Welles et al. (2014). The oxygen consumption rate could be described by Equation 2, which contains a constant factor, representing the residual oxygen consumption rate and a flexible factor, representing the oxygen uptake rate associated with poly-P recovery, growth and glycogen formation. The flexible factor can be described by a similar function as was used for the individual processes.

PAO Aerobic kinetic rates

$$q^o_{O2,PAO\_cor}(S) = q^{0,0}_{O2,PAO\_P,N,Gly} * \frac{1}{1+e^{(2.(S-Si_2))}} + q^{0,0}_{O2,PAO\_res} \qquad \text{(Equation 2)}$$

$$q^o_{PO4,PAO\_cor}(S) = q^{0,0}_{PO4,PAO\_cor} * \frac{1}{1+e^{(bi_3.(S-3))}} \qquad \text{(Equation 3)}$$

$$q^o_{NH4,PAO}(S) = q^{0,0}_{NH4,PAO} * \frac{1}{1+e^{(4.(S-Si_4))}} \qquad \text{(Equation 4)}$$

With

$q^o_{O2,PAO\_cor}(S)$: PAO O_2 uptake rate at different salinity, corrected for maintenance O2 consumption

$q^o_{PO4,PAO\_cor}(S)$: PAO PO_4 uptake rate at different salinity, corrected for maintenance P-release

$q^o_{NH4,PAO}(S)$: PAO NH_4 uptake rate at different salinity

S: Salinity concentration

And fitted parameters,

$q^{0,0}_{O2,PAO\_P,N,Gly}$: PAO O_2 uptake rate at 0% salinity for PO_4, NH_4 and glycogen

bi_2: Impact factor, describing the magnitude of the inhibition effect on the PO_4 uptake rate

Si_2: Salinity concentration at which 50% inhibition of the PO_4 uptake rate occurs

$q^{0,0}_{O2,PAO\_res}$: PAO residual O_2 uptake rate

$q^{0,0}_{PO4,PAO\_cor}$: PAO PO_4 uptake rate at 0% salinity corrected for PO_4 maintenance

bi_3: Impact factor, describing the magnitude of the inhibition effect on the PO_4 uptake rate

Si_3: Salinity concentration at which 50% inhibition of the PO_4 uptake rate occurs

$q^{0,0}_{NH4,PAO}$: PAO NH_4 uptake rate at 0% salinity

bi_4: Impact factor, describing the magnitude of the inhibition effect on the NH_4 uptake rate

Si_4: Salinity concentration at which 50% inhibition of the NH_4 uptake rate occurs

8

General conclusions and outlook

This PhD study can be divided in two separate parts. The first part of the study focused on functional diversity among PAO clades. In the second part of the study the salinity effects on the metabolism of PAO and GAO were assessed. This chapter discusses in more detail the main outcomes of the study, their implications and perspectives for future research.

8.1 Metabolic flexibility and functional diversity among accumulibacter clades

8.1.1 General conclusions

8.1.1.1 Introduction

Recent studies have indicated that different PAO clades may perform functionally differently. For instance Acevedo et al. (2012) suggested that some PAO clades may have the ability to shift their metabolism from a Poly-P dependent metabolism to a glycogen dependent metabolism during short-term experiments when the poly-P content got depleted, while other PAO may not (as efficiently) perform such a metabolic shift. In some other studies (Carvalho et al., 2007; Flowers *et al.*, 2009) it was suggested that PAO I was able to use both nitrite and nitrate as external electron acceptor, while PAO II could only use nitrite as external electron acceptor. However little is known about the functional differences between PAO clades as techniques for identification of the specific clades have been developed relatively recently and no selection strategies have been developed for the selection of specific PAO clades and consequently only few studies were conducted using EBPR cultures highly enriched with specific PAO clades. The main objective of this part of the study was to assess the functional differences among PAO clades regarding the anaerobic metabolism in relation with their storage polymers and to asses if PAO I has the exclusive ability to use nitrate as external electron acceptor. Specific PAO clades were enriched in several SBR operated under different conditions. In short-term experiments, the functional differences between PAO clade I and II anaerobic metabolism in relation with their storage polymers was assessed. In long-term experiments the metabolism of PAO II and its ability to proliferate was assessed at differen P/C influent ratios an in a phosphate limiting system and these results were compared to similar studies conducted with enriched biomass cultures (presumably) dominated with PAO I. In addition, the denitrification pathways (nitrite and nitrate) were assessed of a highly enriched PAO I culture fed with different carbon sources and these results were compared to results of previous studies to verify if PAO I have the exclusive ability to use nitrate as external electron acceptor.

8.1.1.2 Functional diversity of the anaerobic metabolism among PAO clades

In short-term experiments, it was demonstrated that functional differences exist in the anaerobic metabolism of Candidatus Accumulibacter phosphatis clade I and II. Both PAO clades were able to gradually shift their metabolism from a PAO or mixed PAO-GAO metabolism to a GAO metabolism when the biomass poly-P content decreased. However when the P-content was depleted, PAO II performed a GAO metabolism with a significantly faster HAc-uptake rate than PAO I. At high poly-P content PAO I performed a typical PAO metabolism with an anaerobic P-release/HAc-uptake ratio of +/-0.5 P-mol/C-mol, while PAO II performed a mixed PAO-GAO metabolism where still a significant amount of energy for HAc-storage is produced by the intracellular conversion of glycogen into PHA. A comparison of the long-term results to literature suggested that also at long-term performance, functional differences exist among PAO I and PAO II, where PAO II has the ability to proliferate in phosphate limiting systems for many SRT unlike PAO I which was shown in a previous study to wash out from the system at influent phosphate concentrations that are just sufficient for active biomass synthesis.

8.1.1.3 Effect of storage polymers on the metabolism of PAO I and PAO II

This study investigated the relationship between the storage polymers, the anaerobic kinetics rates and the stoichiometry of a highly enriched culture of '*Candidatus* Accumulibacter phosphatis' Clade II and compared the results to those of previous study by Schuler and Jenkins (2003), which based on the stoichiometry seems to be conducted with a PAO I dominated culture. In this study and in the study by Schuler and Jenkins (2003), it was clearly demonstrated the anaerobic metabolism is flexible and can change dependent on the storage polymers. In this study, at higher P-contents, the kinetic P-release rates for HAc-uptake and maintenance were significantly higher, which may have led to higher intracellular ATP/ADP ratios. Consequently, the HAc-uptake rates increased up to an optimal poly-P/glycogen ratio of 0.3 P-mol/C-mol. Above that optimal ratio, the HAc-uptake rate decreased. The changes in the HAc-uptake rates suggest that at low poly-P contents the ATP formation rate is the rate limiting step, while at high P-contents (and, thus, low glycogen contents) the NADH production rate becomes the rate limiting step for HAc-uptake.

8.1.1.4 Denitrification pathways of PAO I

Regarding the functional differences in the denitrification properties among PAO clades, it was suggested in several previous studies that PAO I was able to use nitrate and nitrite as external electron acceptor, while PAO II was only able to use nitrite as external electron acceptor. This study suggests that, PAO I is not able to use nitrate as external electron acceptor, at least not to the extent that enough energy can be generated for phosphate uptake and subsequent poly-P formation in anaerobic/anoxic/oxic systems. Therefore, this study suggests that no functional difference between PAO I and II regarding the ability to use nitrate as external electron acceptor may not exist. However, it cannot be discarded that functional differences still exist at the sub-clade level or that the differences only appear when biomass is cultivated in anaerobic/anoxic systems.

8.1.2. Outlook

8.1.2.1 Functional diversity of the anaerobic metabolism among PAO clades

These findings are of significance because they provide important insights into the functional diversity of *Accumulibacter* clades, regarding their anaerobic metabolism and their physiological response to changes in the phosphate availability from their environment, as well as their ability to use nitrate as external electron acceptor. These findings also contribute to a better understanding of *Accumulibacter* clades population dynamics observed in previous studies, contribute to explain and clarify the controversy concerning the different stoichiometric and kinetic values observed in EBPR systems and are relevant for the development of operational guidelines for combined chemical and biological phosphate removal processes. The results also suggest that the EBPR performance in activated sludge systems may vary dependent on the PAO clades present in the system.

To obtain a full understanding of the functional diversity, further research should focus on all the metabolic processes including the maintenance in the anaerobic phase, poly-P formation, glycogen formation, growth and maintenance in the anoxic (NO_3, NO_2) and aerobic phase. Although the findings of this study demonstrated the functional differences between PAO I and II, a deeper understanding about the metabolic pathways that underlay these functional differences is still lacking. Clarification of the biochemical pathways through genomics and/or proteomics approaches in combination with metabolomics studies using cultures that are highly enriched with specific PAO clades could provide a deeper understanding of the functional differences among PAO clades and may eventually also provide a better understanding about the advantage of having PAO clades with specific biochemical pathways to proliferate under specific environmental conditions.

Considering practical applications, a complete understanding of the PAO clade differentiation would help to clarify if certain PAO clades may be associated with good P-removal performance and others with bad P-removal performance. In this study all PAO clades enriched, performed good P-removal. However it may be that certain PAO clades perform better under specific operational conditions (such as pH, temp. carbon source) or are more robust to specific stress conditions such as salinity or temporal phosphate limitation. For future research it would be good to consider that functional differences may even exist at the sub-clade level.

8.1.2.2. Effect of storage polymers on the metabolism of PAO I and PAO II

This study clearly demonstrated the influence of the storage polymer on the kinetics and stoichiometry of the anaerobic VFA uptake metabolism and based on the observations a regulation mechanisms was proposed based on the intracellular ratios of the conserved moieties. To obtain better understanding how the flexible metabolism is regulated, it would be interesting to conduct metabolic flux studies and investigate how the intracellular ratio of the conserved moieties (ATP/ADP, NADH/NAD+, NADPH/NADP+) are affected by the storage polymer contents under anaerobic, anoxic and aerobic conditions and how these conserved moiety ratios affect the metabolic fluxes of the main biochemical pathways. Considering the functional differences among PAO clades, such studies would need to be performed using biomass that is highly enriched with specific PAO or GAO clades.

The stoichiometry of the anaerobic biochemical VFA conversions has often been considered as an indicator for the relative fractions of PAO and GAO, assuming that both organisms used fixed metabolic pathways. A few recent studies already indicated that at least some PAO clades are able to shift their metabolism from a poly-P dependent to a glycogen dependent metabolism when the poly-P content decreased, leading to a range of different stoichiometric values for highly enriched PAO cultures. In this study it was shown that both PAO I and II have a flexible metabolism, leading to a wide range of different kinetic rates and stoichiometric values. In addition, it was demonstrated that functional differences among PAO clade I and II lead to significant variety in kinetic and stoichiometric properties of highly enriched PAO cultures. These findings significantly change the interpretation of the stoichiometry observed in EBPR systems. If in a particular system (with controlled pH and VFA feed) the anaerobic P-release/HAc-uptake stoichiometry increases over time, this observation used to be explained by an increase in the PAO fraction of the sludge. However, the new insights from this study have shown that such an observation may as well be explained by a microbial shift from PAO II to PAO I or a metabolic shift caused by an increase in the sludge poly-P/glycogen ratio. Similarly, a decrease in the P-release/VFA-uptake ratio may be explained by a decrease in the P-content or a shift from a PAO I to a PAO II culture. This implies that systems that have been exposed to limiting phosphate concentrations due to overdosing of iron or fluctuations in the phosphate concentrations of industrial waste streams, may show a negligible P-release/VFA-uptake (for instance 0.01 P-mol-C-mol) stoichiometry may still contain significant fractions of PAO that have (nearly) depleted poly-P pools, but are still able to remove excessive amount of phosphate from the influent when suddenly exposed to high dissolved phosphate concentrations. Alternatively, systems that have been exposed to high phosphate concentrations for instance due to secondary P-release in the sludge line of activated sludge systems in the summer period may show a very high P-release/VFA-uptake (for instance 0.70 P-mol-C-mol) stoichiometry, while containing similar fractions of PAO that are saturated with poly-P and therefore are not able to remove more dissolved phosphate from the influent.

8.1.2.3 Denitrification pathways of PAO I

The findings of this study suggest that 'Candidatus Accumulibacter phosphatis' clade IA is not capable of using nitrate as external electron acceptor, suggesting that either the essential genes are not present or that essential genes are not expressed under the operational conditions of this study, in particular A_2O conditions. In a few metagenomic studies by Martin et al. (2006), Flowers et al. (2013), Mao et al. (2014) and Skennerton et al. (2014) in total eleven different 'Candidatus Accumulibacter phosphatis' metagenomes were obtained from 5 different PAO clades (IA, IB, IC, IIA, IIC, IIF). In these studies it was shown that the metagenome of only PAO clade IIF encodes the nitrate reductase and nitrite reductase genes, while the metagenomes of the other clades (IA, IB, IC, IIA, IIF) encoded genes for periplasmic nitrate reductase and nitrite reductase. Periplasmic nitrate reductases plays different physiological roles in different type of organisms and it was observed in previous studies that the activity of periplasmic nitrate reductase was not sufficient to support anoxic respiration in several microorganisms (Skennerton et al., 2014; Moreno-Vivian et al., 1999).

Although highly enriched cultures were obtained in this research, the study on the denitrification pathways in an A_2O system, suggests that the presence of side populations cannot be avoided. Moreover, a literature review presented in this study indicated that these side populations may play a crucial role in the metabolic conversions observed in many past studies with A_2O system, which was further supported by basic calculations that demonstrated the potential role of these side populations on the denitrification activities. To study the metabolic properties of specific PAO and GAO clades, without interference of activity from side populations, it would be very useful to work with pure cultures. Although many past studies aimed at isolation of PAO and GAO clades, pure cultures have not been obtained yet. Therefore, it would be interesting to develop new strategies for the isolation of PAO and GAO clades.

8.1.2.4 Metabolic models
Significant functional differences have been observed between PAO clade I and II. Under similar conditions, but dependent on the polymer contents, a four times difference may exist in the kinetic rates of PAO I and II and a two times difference may exist in the stoichiometry of the anaerobic conversions of PAO I and II. In addition a wide range of kinetic and stoichiometric parameters can be observed for each specific PAO clade due to the metabolic flexibility in relation with the storage polymer contents. These differences urge the need to develop PAO clade specific EBPR models which include the metabolic flexibility. This would lead to metabolic models that describe better the metabolic conversions and may also lead to a better understanding and prediction of process performance in particular in wastewater treatment systems that perform under dynamic conditions for instance for industrial applications.

8.1.2.5 Metagenomic analysis of PAO and GAO clades
To explore the functional differences among different PAO clades, and to develop genome based metabolic models, it would be very useful to have reliable and high quality (meta)genomes obtained from highly enriched or isolated cultures of specific PAO and GAO clades. Currently eleven different *Accumulibacter* genomes (Martin et al., 2006; Flowers et al., 2013; Skennerton et al., 2014; Mao et al., 2014) 2 *Competibacter* genomes (McIlroy et al., 2013) and two *Defluviicoccus* genomes (Wang et al., 2014; Nobu et al., 2014) exist. However, most of these genomes were obtained through (deep) metagenomics approaches using biomass samples from which the fractions of specific PAO clades were either not reported (9 metagenomes, Skennerton et al., 2014, Mao et al., 2014) or were not highly enriched, i.e. containing 74% Clade IIA and 7% Clade IA (culture R104-IIA) and containing 4% Clade IIA and 36% Clade IA (culture R107-IA) (2 metagenomes, Martin et al., 2006; Flowers et al., 2013). The existing genomes still contain gaps (Skennerton et al., 2014). Furthemore, considering the fact that the core metabolism of PAO and GAO clades is very comparable to that of ordinary heterotrophic bacteria and in particular closely related PAO and GAO clades, there may be very conserved domains in the genomes among different bacteria, implying that there is a significant chance for the inclusion of foreign genes into the PAO or GAO genomes (Skennerton et al., 2014; Flowers et al., 2013). In addition, it was

demonstrated in the present study by DGGE analysis, that DNA extracted from highly enriched PAO cultures (i.e. 99% PAO I/EUB), still contained significant fractions of other micro-organisms such as *Bacteroidetes*. This may be caused by differences in the extraction efficiency, which is already well known for *Defluviicoccus*, and by differences in the cell volume of different type of bacteria, affecting the DNA content per dry weight biomass. To obtain more reliable and complete genomes, it would be recommendable that future genomics studies are conducted with highly enriched or even isolated PAO and GAO cultures. Finally it would also be more useful, if future genomics studies are combined with transcriptomic, proteomic or physiology studies that can confirm the presence or absence of specific metabolic activities that are explored in the metagenomes as was done by Mao et al. (2014).

8.1.2.6 Selection methods and isolation of specific PAO and GAO clades

Although the PAO clades were enriched on a trial and error base, the findings of this study provide insight in the operational conditions that may favor the proliferation of specific PAO clades. These insights may help to develop selection strategies for the enrichment of specific PAO clades to enable future research on specific PAO clades. This study suggests that high operational pH, the presence of propionate (a more reduced VFA) and high influent P/C ratios favor the enrichment of PAO I, whereas neutral pH with acetate as carbon source and low influent P/C ratios favor the enrichment of PAO II. Considering that PAO I would have the exclusive ability to use nitrate as external electron acceptor, it was suggested in past studies, that the introduction of an anoxic zone with nitrate in laboratory EBPR systems could serve as a selection method for the development of highly enriched PAO I cultures. The present PhD research suggests however that such an anoxic zone in A_2O systmes with nitrate would not give a competitive advantage to PAO clade I. However, it is still possible that the introduction of an anoxic zone can lead to the enrichement of specific PAO clades. The operation under A_2O conditions applied in this study, still allows the proliferations of PAO that are not capable of using nitrate as external electron acceptor with a higher biomass yield than potential PAO that would be able to use nitrate as external electron acceptor. Therefore, to support the development of PAO that are capable of using nitrate as external electron acceptor, more selective conditions should be applied, i.e. A_2 conditions, that only allow the proliferation of potential PAO capable of using nitrate as external electron acceptor. More enrichment studies are needed to confirm and optimize the conditions that lead to the enrichment of specific PAO clades and avoid the prevalence of GAO under low influent P/C ratios. The development of successful and reliable enrichment strategies for specific PAO clades will support future research on the clade differentiation and it will also help to find more consistency between different EBPR studies, eliminating the possibility to use the PAO clades differentiation as a scapegoat for lack of reproducibility.

Although this PhD research has led to more insight in the conditions that may favor the proliferation of specific PAO clades, it was also demonstrated that limitation of phosphate, a selection method which has often been used in previous studies for the enrichment of GAO, is unreliable and may still allow the prevalence of significant fractions of PAO II. To support research on PAO-GAO competition, it would be useful to develop more reliable GAO enrichment strategies.

8.2 Salinity effects on PAO and GAO

8.2.1 General conclusions

8.2.1.1 Introduction

When saline water is used as secondary quality water, the generated wastewater can be collected by a combined sewerage network, or the black water can be collected using a separate collection system. This practice will lead to rise in salinity of the wastewater. The salinity in the saline water source, the percentage of fresh water replacement and the design of the collection system affect the salinity concentration in the saline wastewater, leading to a wide range of possible salinity concentrations in the wastewater ranging from 0% to 4.1% salinity. Separate collection of black water requires a dual collection network, but will in turn allow for the potential reuse of non-saline wastewater. The salinity concentrations in industrial wastewater streams also vary over a wide range from 0% salinity up to 15% salinity (Lefebvre et al., 2006). When the saline water is treated in wastewater treatment plants, the salinity in the wastewater may create deleterious effects to the microorganisms present in the activated sludge and, thereby, affecting the COD, N and P removal processes in general. The main objective of this part of the study was to get a better understanding about the effect of salinity on the metabolism of the microbial populations that prevail in the enhanced biological phosphorus removal process (PAO and GAO) to support the development of design and operational guidelines for treatment plants treating saline wastewater generated when saline water is used as secondary quality water in the urban water cycle, when saline water intrusion in the sewerage takes place or when generated by industry.

8.2.1.2 Short-term salinity effects on the anaerobic metabolism of PAO and GAO

It was demonstrated in short-term tests that salinity affected the anaerobic metabolism of both PAO and GAO, with PAO being the more sensitive organisms. With increasing salinity the HAc uptake rates were inhibited while the maintenance requirements increased (up to 4% salinity) for both PAO and GAO. Interestingly, elevated salinity levels seemed to induce a metabolic shift from a Poly-P dependent metabolism towards a glycogen dependent metabolism for both anaerobic VFA uptake and maintenance activity.

8.2.1.2 Short-term salinity effects on the aerobic metabolism of PAO and GAO

The aerobic metabolism of PAO was more severally affected by salinity than the anaerobic metabolism. An increase from 0.02 to 0.18% salinity led to a decrease in the specific oxygen consumption, PO_4 and NH_4 uptake rates of 25%, 46% and 63%, respectively. At 0.35% and higher salinity concentrations, the PO_4-uptake, NH_4-uptake and glycogen recovery were fully inhibited. Similar to the anaerobic metabolism, salinity seemed to inhibit the different metabolic activities differently, leading to a metabolic shift.

8.2.1.3 Quantitative description of salinity effects

As the different metabolic processes were affected by salinity differently, equations were developed, describing the salinity effects on the initial kinetic rates of the different metabolic processes in the anaerobic and aerobic phase of PAO and GAO (only the anaerobic metabolism). The equations were constructed partially on the basis of assumptions regarding the mechanisms of salinity effects and partially using empirical equations. The resulting model could describe the salinity effects on the different metabolic processes of PAO and GAO (only the anaerobic metabolism).

8.2.2 Outlook

8.2.2.1 Salinity effects on the metabolism of PAO and GAO

The findings in this PhD research suggest that any saline discharge equivalent to more than 5% seawater (with 3.4% salinity) or 15% brackish water (with 1.2% salinity) by either seawater toilet flushing, industrial discharges or saline intrusion can cause serious upsets of the EBPR process. However, the presented data is based on short-term (hours) experiments where PAO and GAO were exposed to sodium chloride, without giving the organisms the opportunity to acclimatize to higher salinity concentrations or giving the system the possibility to select for more salt tolerant PAO strains. In this respect, Bassin et al. (2011) and Pronk et al. (2013) demonstrated in two studies with granular sludge, that after a long-term acclimatization to elevated salinity concentrations, PAO were able to perform EBPR activity at salinity concentrations up to 2.2% salinity. Studies focused on the long-term salinity effects on EBPR cultures should also consider the composition of the synthetic wastewater as well as the salt composition as these may affect the acclimatization process. It is well known bacteria originating from fresh water conditions, often accumulate compatible solutes intracellularly to balance the osmotic pressure in saline environments (Brown, 1976, Galinski and Truper, 1994). Some bacteria are able to synthesize those compatible solutes but other organisms acquire tolerance to salinity based on the accumulation of compatible solutes from the environment (Jebbar et al., 1992). This implies that availability of compatible solutes or precursors of compatible solutes in (synthetic) wastewater (for example glycine betaine from yeast extract) could support the tolerance to salinity of certain micro-organisms (Galinski and Truper, 1994). The main constituents of seawater are sodium and chloride ions, but seawater also contains other ions such as sulphate, magnesium, calcium, potassium and others. In this study only sodium chloride was used to avoid potential phosphate precipitations with magnesium and calcium and to avoid potential sulphate reduction by sulphate reducing bacteria. However, most saline wastewater will probably have a composition that is similar to that of seawater and this composition may also affect the tolerance to salinity. In particular the potassium/sodium ratio may affect the salinity tolerance of bacteria as potassium is an important signaling ion that triggers the production/accumulation of compatible solutes (Roesler et al., 2001) and in addition potassium serves as a counterion of poly-P, which is transported over the membrane during the anaerobic and aerobic phase along with phosphate.

8.2.2.2 Indirect salinity effects on PAO in BNR systems

When wastewater contains salinity, other microbial populations in the sewage or the activated sludge that are responsible for other nutrient removal processes may be affected by the salinity too, leading to the accumulation or production of substances that negatively affect the EBPR process. For instance, potential NO_2 accumulation as a consequence of the possible inhibition of NOB by salts (Vredenbregt *et al.*, 1997; Dincer and Kargi, 1999; Intrasungkha *et al.*, 1999; Wu *et al.*, 2008; Cui *et al.*, 2009) could also directly affect the activity of PAO and GAO and thereby also the competition between PAO and GAO (Pronk et al., 2014). Finally, SO_4 (abundant in seawater) can be converted to H_2S by sulfate reducing bacteria (SRB) under anaerobic conditions in the sewage or in thick biofilms in the treatment systems. H_2S has an inhibiting effect on many microorganisms, in general. Two recent studies (Saad et al., unpublished a, b; Rubio Rincon et al., unpublished) demonstrated that in particular the aerobic metabolism of PAO I is severely affected by H_2S . Further research is needed to assess the salinity effects on the EBPR activity in BNR systems with the aim to find operational parameters that minimize the accumulation/production of inhibiting substances by other microbial communities.

8.3 References

Acevedo B., Oehmen A., Carvalho G., Seco A., Borras L., Barat R., (2012) Metabolic shift of polyphosphate-accumulating organisms with different levels of poly-phosphate storage. Water Research 46, 1889-1900

Brown A.D. (1976) Microbial water stress. Bactcriol. Rev. 41: 803- 846.

Carvalho G., Lemos P.C., Oehmen A., Reis M.A.M. (2007) Denitrifying phosphorus removal: linking the process performance with the microbial community structure. Water Research 41(19), 4383, 4396

Cui Y., Peng C., Peng Y., Ye L. (2009) Effects of Salt on Microbial Populations and Treatment Performance in Purifying Saline Sewage Using the MUCT Process. Clean 37(8): 649-656

Dincer A. R., Kargi F. (1999) Salt Inhibition of Nitrification and Denitrification in Saline. *Environmental Technology* 20(11), 1147-1153

Flowers J.J., He S., Yilmaz S., Noguera D.R., McMahon K.D. (2009) Denitrification capabilities of two biological phosphorus removal sludges dominated by different 'Candidatus *Accumulibacter*' clades. Environmental Mivrobiology Reports 1(6): 583-588

Flowers, J. J., He, S., Malfatti, S., del Rio, T. G., Tringe, S. G., Hugenholtz, P., & McMahon, K. D. (2013). Comparative genomics of two 'Candidatus Accumulibacter'clades performing biological phosphorus removal. *The ISME journal*, *7*(12), 2301-2314.

Galinski E.A., Truper H.G. (1994) Microbial behavior in salt-stressed ecosystems. FEMS Microbiology Reviews 15: 95-108

Intrasungkha N., Keller J., Blackall L.L. (1999) Biological nutrient removal efficiency in treatment of saline wastewater. Water Science and Technology 39(6): 183-190

Jebbar M., Talibart R., Gloux K., Bernard T., Blanco C. (1992) Osmoprotection of Escherichia coli by ectoine: uptake and accumulation characteristics. J. Bacteriol. 174: 5027-5035.

Lefebvre O., Moletta R., (2006) Treatment of organic pollution in industrial saline wastewater: A literature review. Water Research 40: 3671-3682.

Mao, Y., Yu, K., Xia, Y., Chao, Y., & Zhang, T. (2014). Genome reconstruction and gene expression of "Candidatus Accumulibacter phosphatis" clade IB performing biological phosphorus removal. *Environmental science & technology*, *48*(17), 10363-10371.

Martín, H. G., Ivanova, N., Kunin, V., Warnecke, F., Barry, K. W., McHardy, A. C., Yeates C., He S., Salamov A.A., Szeto E., Dalin E., Putnam N.H., Shapiro H.J., Pangilinan J.L., Rigoutsos I., Kyrpides N.C., Blackall L.L., McMahon K.D., Hugenholtz, P. (2006). Metagenomic analysis of two enhanced biological phosphorus removal (EBPR) sludge communities. *Nature biotechnology*, *24*(10), 1263-1269.

McIlroy, S. J., Albertsen, M., Andresen, E. K., Saunders, A. M., Kristiansen, R., Stokholm-Bjerregaard, M., Nielsen K.L., Nielsen, P. H. (2014). 'Candidatus Competibacter'-lineage genomes retrieved from metagenomes reveal functional metabolic diversity. *The ISME journal*, *8*(3), 613-624.

Moreno-Vivián, C., Cabello, P., Martínez-Luque, M., Blasco, R., & Castillo, F. (1999). Prokaryotic nitrate reduction: molecular properties and functional distinction among bacterial nitrate reductases. *Journal of bacteriology*, *181*(21), 6573-6584.

Nobu, M. K., Tamaki, H., Kubota, K., Liu, W. T. (2014). Metagenomic characterization of 'Candidatus Defluviicoccus tetraformis strain TFO71', a tetrad-forming organism, predominant in an anaerobic–aerobic membrane bioreactor with deteriorated biological phosphorus removal. *Environmental microbiology*, *16*(9), 2739-2751.

Pronk M., Bassin J.P., de Kreuk M.K., Kleerebezem R., van Loosdrecht M.C.M. (2013) Evaluating the main side effects of high salinity on aerobic granular sludge. Appl Mirobiol Biotechnol 98: 1339-1348

Bassin, J. P., Pronk, M., Muyzer, G., Kleerebezem, R., Dezotti, M., & Van Loosdrecht, M. C. M. (2011). Effect of elevated salt concentrations on the aerobic granular sludge process: linking microbial activity with microbial community structure. Applied and environmental microbiology, 77(22), 7942-7953.

Roesler M., Muller V. (2001) Osmoadaptation in bacteria and archae: common principles and differences. Environmental Microbiology 3(12): 743-754

Saad S., Welles L., Lopez-Vazquez C.M., van Loosdrecht M.C.M., Brdjanovic D. (unpublished a) Impact of sulphide on the anaerobic metabolism of 'Candidatus Accumulibacter phosphatis' clade I.

Saad S., Welles L., Lopez-Vazquez C.M., van Loosdrecht M.C.M., Brdjanovic D. (unpublished b) Impact of sulphide on the aerobic metabolism of 'Candidatus Accumulibacter phosphatis' clade I.

Skennerton, C. T., Barr, J. J., Slater, F. R., Bond, P. L., & Tyson, G. W. (2014). Expanding our view of genomic diversity in Candidatus Accumulibacter clades. Environmental microbiology 17(5), 1574-1585

Rubio Rincon F. J., Welles L., Lopez-Vazquez C.M., van Loosdrecht M.C.M., Brdjanovic D. (unpublished) Sulphide inhibition on the metabolism of 'Candidatus Accumulibacter phosphatis' clade I.

Schuler, A.J., and Jenkins, D. (2003) Enhanced Biological Phosphorus Removal from Wastewater by Biomass with Different Phosphorus Contents, Part 1: Experimental Results and Comparison with Metabolic Models. Water Environ Res 75(6): 485-498

Vredenbregt L.H.J., Potma A.A., Nielsen K., Kristensen G.H., Sund C. (1997) Fluid bed biological nitrification and denitrification in high salinity wastewater. Water Sci. Technol. 36 (1), 93–100

Wang, Z., Guo, F., Mao, Y., Xia, Y., Zhang, T. (2014). Metabolic Characteristics of a Glycogen-Accumulating Organism in Defluviicoccus Cluster II Revealed by Comparative Genomics. Microbial ecology, 68(4), 716-728.

Wu G., Guan Y., Zhan X. (2008) Effect of salinity on the activity, settling and microbial community of activated sludge in a sequencing batch reactors treating synthetic saline wastewater. Water Science and Technology 58(2): 351-358

Laurens Welles was born in 1983 in The Hague, the Netherlands. During his high school time at Christelijk College de Populier, his interest was in the natural sciences. He was very fascinated by the concept of evolution and eager to learn how things around him were working. Once he graduated from high school, he left to Australia (and later Indonesia) where he spent his time by working and travelling. After a year of adventure and many hard labour jobs, he came back to the Netherlands to study.

In 2002, he started a BSc study in Life Science and Technology at the Delft University of Technology and Leiden University. During this study, he became very interested in micro-organisms. He continued at Delft University of Technology in 2006 with a MSc. study in Life Science and Technology, specialization 'Cell Factory', in which he got really fascinated by applications that make use of mixed microbial populations for conversion or recovery of nutrients. To further explore his interest in these applications, he conducted his MSc research project on the Enhanced Biological Phosphorus Removal process at the University of Tokyo followed by an internship focused on bioremediation at Kurita Water Industries in Japan. In 2010, he was appointed as PhD fellow in the SALINE project at UNESCO-IHE and consortium partners Delft University of Technology, KWR Watercycle Research Institute, University of Cape Town, The Hong Kong University of Science and Technology, The Higher Polytechnic Institute "José Antonio Echeverría,"and Birzeit University.

Currently Laurens is still interested in the development and implementation of bioprocess technologies for nutrient removal and recovery from waste. Since 2015, Laurens is working as a postdoctoral researcher for UNESCO-IHE and Delft University of Technology. At UNESCO-IHE he assists in the management of the project, "Stimulating local innovation in sanitation for Sub-Saharan Africa and South East Asia" funded by the Bill & Melinda Gates. At the Delft University of Technology he conducts fundamental research on the niche differentiation of microbial communities in EBPR systems, in the SIAM project, "microbes for health and environment", funded by the Gravitation programme of the Dutch Ministry of Education, Culture and Science. In addition, Laurens works as a freelance consultant and executes optimization projects for nutrient removal processes in wastewater treatment plants in the Netherlands, commissioned by the water board.

List of publications

Conference proceedings

Welles, L., Lopez-Vazquez, C. M., van Loosdrecht, M. C. M., & Brdjanovic, D. (2012). Use of saline water in the urban water cycle to alleviate fresh water stress: a feasibility assessment for biological wastewater treatment. In *IWA World Water Congress & Exhibition. September 16–21st, 2012, Busan, Korea.*

Saad, S. A., Welles, L., Lopez-Vazquez, C. M., van Loosdrecht, M. C. M., & Brdjanovic, D. (2013). Sulfide effects on the anaerobic kinetics of phosphorus-accumulating organisms. In *13th World Congress on Anaerobic Digestion. June 25-28th, 2013, Santiago de Compostela, Spain.*

Journal articles

Welles, L., Lopez-Vazquez, C. M., Hooijmans, C. M., Van Loosdrecht, M. C. M., & Brdjanovic, D. (2014). Impact of salinity on the anaerobic metabolism of phosphate-accumulating organisms (PAO) and glycogen-accumulating organisms (GAO). *Applied microbiology and biotechnology, 98*(17), 7609-7622.

Welles, L., Lopez-Vazquez, C. M., Hooijmans, C. M., van Loosdrecht, M. C. M., & Brdjanovic, D. (2015). Impact of salinity on the aerobic metabolism of phosphate-accumulating organisms. *Applied microbiology and biotechnology, 99*(8), 3659-3672.

Welles, L., Tian, W. D., Saad, S., Abbas, B., Lopez-Vazquez, C. M., Hooijmans, C. M., van Loosdrecht M.C.M., Brdjanovic, D. (2015). Accumulibacter clades Type I and II performing kinetically different glycogen-accumulating organisms metabolisms for anaerobic substrate uptake. *Water research, 83,* 354-366.

Articles under review or in prepration

Welles, L., Abbas, B., Sorokin, D.Y., Lopez-Vazquez, C. M., Hooijmans, C. M., van Loosdrecht M.C.M., Brdjanovic, D. (submitted). Metabolic response of '*Candidatus* Accumulibacter phosphatis' clade II to changes in the influent P/C ratio. *Water research*

Welles, L., Lopez-Vazquez, C. M., Hooijmans, C. M., van Loosdrecht M.C.M., Brdjanovic, D. (submitted). Prevalence of 'Candidatus Accumulibacter phosphatis' clade II under phosphate limiting conditions. *Applied Microbiology and Biotechnology*

Saad, S., Welles, L., Abbas, B., Lopez-Vazquez, M.C, van Loosdrecht, M.C.M., Brdjanovic, D. (submitted) Denitrification pathways of PAO clade I with different carbon sources. *Water Research*

Welles, L., Abbas, B., Lopez-Vazquez, C. M., Hooijmans, C. M., van Loosdrecht M.C.M., Brdjanovic, D. (in preparation). Factors affecting the salinity tolerance of '*Candidatus* Accumulibacter phosphatis' clades during long-term exposure to salinity.

Welles, L., Abbas, B., Lopez-Vazquez, C. M., Hooijmans, C. M., van Loosdrecht M.C.M., Brdjanovic, D. (in preparation). Impact of seasalt on microbial community structure and metabolism of an enriched EBPR culture during long-term exposure.

Saad, S., Welles, L., Abbas, B., Lopez-Vazquez, M.C, van Loosdrecht, M.C.M., Brdjanovic, D. (in preparation) Sulfide effects on the anaerobic metabolism of phosphate-accumulating organisms.

Saad, S., Welles, L., Abbas, B., Lopez-Vazquez, M.C, van Loosdrecht, M.C.M., Brdjanovic, D. (in preparation) Sulfide effects on the aerobic metabolism of phosphate-accumulating organisms.

Acknowledgements

This PhD research study was carried out as part of the SALINE project led by UNESCO-IHE and consortium partners Delft University of Technology, KWR Watercycle Research Institute, University of Cape Town, The Hong Kong University of Science and Technology, The Higher Polytechnic Institute "José Antonio Echeverría," and Birzeit University. The SALINE project was financed by UNESCO-IHE internal research fund with a special generous contribution from Professor George Ekama from University of Cape Town. The main objective of the project was to assess the feasibility of using saline water directly for non-potable purposes such as flushing toilets in urban areas to mitigate fresh water consumption. Within this project, the aim of his PhD research was to get a better understanding of the salinity effects on polyphosphate-accumulating organisms (PAO) and glycogen accumulating organisms (GAO) in the Enhanced Biological Phosphorus Removal (EBPR) process in activated sludge systems to support the design and development of operational guidelines for WWTPs treating saline wastewater generated when saline water is used directly for non-potable purposes. For this purpose, it was planned to enrich a PAO and a GAO cultures under fresh water conditions, which then could be exposed in short and long term experiments to elevated salinity levels.

Due to 'interesting' observations, the course of the research project deviated a lot from the initial plans. After a very long period of dedication to my reactors and passion for my bacteria, I obtained in my PAO reactor an enrichment culture that, from the anaerobic metabolic conversions, seemed to be a GAO dominated culture. In the GAO reactor I obtained an enrichment culture that from the metabolic conversions seemed to be a GAO culture, but a FISH microscopic analysis showed that it was a PAO dominated culture. In conclusion, it seemed that I had GAO in my PAO reactor and strange PAO in my GAO reactor. This was one of the 'exciting' moments of my PhD. Additional FISH analysis on the biomass culture in the PAO reactor clarified that this culture was actually a PAO II dominated culture that performed a partial GAO metabolism. What a relief, now I only had one strange PAO culture in two different reactors. In the mean time I had some talks with Tian Wende (a colleague from TUDelft) who seemed to have similar observations with his mixed PAO I-II enrichment culture and Sondos Abdel Hakeem Abdel Aal Saad (my colleague from UNESCO-IHE), who enriched a PAO I dominated culture that showed characteristics, which were comparable to those observed in past PAO studies in Delft. These observations led to a new research plan focused on the functional diversity of PAO clades. Finally both the fundamental research on the functional diversity of PAO clades and the original saline research were satisfactorily completed, although not all saline studies (long-term experiments) are included in this thesis. So, after several years of work with the PAO and GAO, this project has come to an end. It has been an interesting journey, with hard work, stress, many PAO and GAO dreams, sometimes disappointment which was often followed by satisfaction in a later stage.

First of all, I would like to express my gratitude to my promotors Damir Brdjanovic and Mark van Loosdrecht, for giving me the unique opportunity to do this PhD research and for their support during the project. Both of them shared their knowledge and gave me very valuable advice of different nature in different stages of my research, where Damir often commented on my research from the perspective of practical applications, while Mark provided me usually with feed-back on the fundamental aspects of my research. I very much appreciated the freedom that was given to me, to deviate from the original research plans and focus on the functional diversity of the PAO clades, before coming back to the saline research.

Secondly, I would like to express my thankfulness to my mentors Carlos Lopez-Vazquez and Tineke Hooijmans for their valuable input and support during this project. In the meetings, we had nice discussions, where Carlos shared his broad knowledge and expertise in the field of EBPR and Tineke asked me the critical questions that forced me to reconsider my research plans. Together they were a perfect combination of mentors and I have learnt many things from them.

Besides my promotors and mentors, there are other people who have contributed to my research directly or indirectly. Thanks to Yang Jiang and Leonie Marang who taught me the analytical procedures for PHA analysis and Mario Pronk for the many PAO-GAO talks. Thanks to Tian Wende and Sondos Saad for their collaboration in the research on PAO I and II and to Yumei Lin for her collaboration on the characterization of PAO ALE. We have spent quite some time in the laboratory together doing experiments and discussing about the findings. I really enjoyed working with them. Thanks to Tessa van den Brand, who also conducted her PhD research as part of the SALINE project, for the nice chats and discussions about our research on saline wastewater treatment and other things during conferences and after project meetings. Thanks to Dimitri Yukatan Sorokin for his contribution in this research. Once he introduced me during my Bachelor studies to the world of the microorganisms and during this research project he continued having inspiring discussions with me and showing me the beauty of the microorganisms. Also, thanks to Robbert Kleerebezem, Jelmer Tamis and Steef de Valk for their valuable lunchtime discussions, sometimes from my side on the PAO metabolism, and other times on potential remarkable applications of biodegradable plastics, where Steef often took the lead in the discussions. Once in a while Moustafa Moussa, who is an expert on saline wastewater treatment, appeared in UNESCO-IHE. He always showed great interest in my work and shared his knowledge with me, which was highly appreciated.

One of the great things in UNESCO-IHE is the international environment in which I, myself being Dutch, seemed to be an exotic minority. I really enjoyed to spent time with my non-exotic colleagues from all over the world in the lab and in the office. My first office members were Saeed Baghoth, Loreen Villacorte, Chol Deng Thong Abel and Salifu Abdulai who made me feel at home. Later I moved to a different office in the Oude Delft, meant for the sanitary engineering students (sanitary dreams office). In there, my first office and lab member was Javier Sanchez-Guillen from Panama. We have had a great time together, in particular on the days that we were running experiments from the early hours until the security forced us out of the building. On those days, when our family was in Panama and Korea, we sometimes ended the day in somehow, eating food somewhere, often with someone else that left the building late, like Salifu Abdulai who usually also enjoyed the research until the late hours. The next sanitary dreams office member was Peter Mawioo from Kenya, followed by Fiona Zakaria from Indonesia and Joy Riungu from Kenya. You can see that in the office there were already more Kenyans than Dutch. In a later stage Sondos Saad from Egypt became member of the office, followed by Sangyeob Kim (Mr. Kim) from South Korea and Yuli Ekowati from Indonesia (second Indonesian). The latest member who joined the team was Francisco Rubio Rincon (Pancho) from Mexic. Currently he is having a great time with the PAO, SRB and possibly many other exciting microorganisms, which we will see in near future. Besides my office members I have had nice interactions and became good friends with many other PhD colleagues from UNESCO-IHE and TUDelft (too many too list here). Although the major part of my research was conducted in UNESCO-IHE, I visited TUDelft now and then to conduct my analytical work. Despite the relative few moments that I was in TUDelft, I enjoyed very much the time in there and felt like a full member of the EBT group. Thanks to all of them.

Special thanks to the UNESCO-IHE laboratory staff, Fred Kruis, Frank Wiegman, Peter Heerings, Ferdi Battes, Lyzette Robbemont, Berend Lolkema and Don van Galen, for all their support during the research project and to the TUDelft laboratory staff, Ben Abbas and Mitchel Geleijnse for their contribution to this research with the molecular work, the technical discussions on how to get the samples loaded in the denaturing gradient gels during the lunch times with 'broodje flaplap'. Thanks to Udo van Dongen for helping me with FISH-ing. Thanks to my students, Carmen Almeyda, Rebecca de Vera, Don van Galen and Malan Kandage Jeevika Prabodini for their efforts and contributions and to many other students that contributed to the good environment in the sanitary engineering lab.

Finally I would like to express my gratitude to my family. Many thanks to my parents, Ineke Zwaan and Harrie Welles, who supported me a lot, especially in the beginning of my PhD and in a later stage when my son got born. Thanks to my brother, Fokke Welles, his wife, Xaviera Jahier, my beautiful niece, Sarah Welles and wonderful nephew, Rezah Welles, for their help in difficult times. Eventhough my parents-in-law and brother-in-law live on the other side of the world, they have made great efforts as well to support me during my PhD project, in particular when I visited them after my conferences. The intensive acupuncture treatment and ginseng extracts that I received were very effective. 아버님(허태), 어머님(이원남), 처남(허수만) 저희를 도와 주시고, 응원 해 주셔서 감사합니다. 앞으로도 더 열심히 노력하는 모습 보여 드리도록 하겠습니다. Last, but most important, thanks to my lovely fiancé, Jahmahn Hur, and my wonderful son, Elias Gangsan Welles who got born in the third year of my PhD. Without their love and continuous support I would not have been able to bring this project to a successful end. 사랑해!

T - #0401 - 101024 - C24 - 244/170/13 - PB - 9781138029477 - Gloss Lamination